KW-484-055

PERGAMON INTERNATIONAL LIBRARY
of Science, Technology, Engineering and Social Studies
The 1000-volume original paperback library in aid of education,
industrial training and the enjoyment of leisure

Publisher: Robert Maxwell, M.C.

An Introduction to

TURBULENCE AND ITS MEASUREMENT

SOME OTHER BOOKS IN THIS SERIES

BENHAM
Elementary Mechanics of Solids

BENSON
Advanced Engineering Thermodynamics

BRADSHAW
Experimental Fluid Mechanics 2nd Edition

BUCKINGHAM
The Laws and Applications of Thermodynamics

DIXON
Fluid Mechanics, Thermodynamics of Turbomachinery
Worked Examples in Turbomachinery (Fluid Mechanics and
Thermodynamics)

MOORE
The Friction and Lubrication of Elastomers

MORRILL
An Introduction to Equilibrium Thermodynamics

PEERLESS
Basic Fluid Mechanics

The terms of our inspection copy service apply to all
the above books. A complete catalogue of all books in
the Pergamon International Library is available on
request.

The Publisher will be pleased to receive suggestions
for revised editions and new titles.

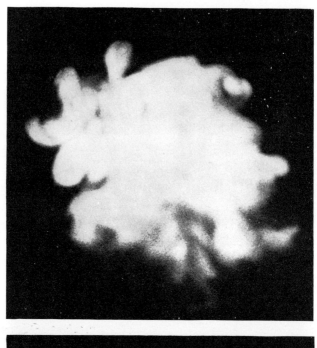

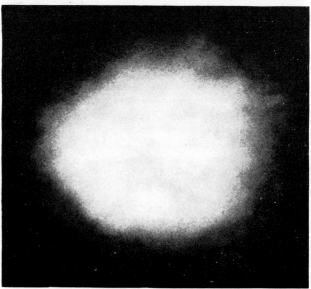

(*Frontispiece*) Illuminated cross-section of a smoke-filled turbulent jet (see Fig. 3). *Top*: short exposure. *Bottom*: long exposure, showing approximate mean concentration. How much would we know about turbulence if we measured only mean quantities?

An Introduction to Turbulence and its Measurement

P. BRADSHAW
Aeronautics Department,
Imperial College of Science and Technology

PERGAMON PRESS
Oxford · New York · Toronto
Sydney · Paris · Braunschweig

U.K.	Pergamon Press Ltd., Headington Hill Hall, Oxford OX3 0BW, England
U.S.A.	Pergamon Press Inc., Maxwell House, Fairview Park, Elmsford, New York 10523, U.S.A.
CANADA	Pergamon of Canada, Ltd., 207 Queen's Quay West, Toronto 1, Canada
AUSTRALIA	Pergamon Press (Aust.) Pty. Ltd., 19a Boundary Street, Rushcutters Bay, N.S.W. 2011, Australia
FRANCE	Pergamon Press SARL, 24 rue des Ecoles, 75240 Paris, Cedex 05, France
WEST GERMANY	Pergamon Press GMbH, D-3300 Braunschweig, Postfach 2923 Burgplatz 1, West Germany

First edition 1971

Reprinted with corrections 1975

Library of Congress Catalog Card No. 75-152527

Printed in Great Britain by A. Wheaton & Company, Exeter
ISBN 0 08 016621 0 (flexicover)
ISBN 0 08 016620 2 (hard cover)

Editorial Introduction

THE books in the Thermodynamics and Fluid Mechanics division of the Commonwealth Library were originally planned as a series for undergraduates, to cover primarily those subjects taught in a three-year undergraduate course for Mechanical Engineers. However it was considered that some of the books should also prove useful to aeronautical engineering undergraduates, and to graduate students.

The present book is somewhat different, in that it should be most useful as an introductory text for post-graduate courses, particularly in aeronautics but also in more general fluid mechanics. However, as the author points out, it should be readily understandable to anybody with a good general background in engineering fluid dynamics.

Contents

Preface

TURBULENCE is the most common, the most important and the most complicated kind of fluid motion. It is peculiarly resistant to mathematical treatment: indeed, turbulence studies may be defined as the art of understanding the Navier–Stokes equations without actually solving them. This may seem to classify the subject as a branch of lifemanship rather than of science, but it is helpful in reminding us that we do possess a closed set of apparently accurate equations which, some day, we may be able to solve numerically for cases of engineering interest. At present we must rely on measurements, either to provide the required information directly or to guide the empirical assumptions necessary in simplified prediction methods. For engineering purposes we usually need to know only mean quantities like pressure drop or surface heat transfer rate, and the assumptions needed in calculation methods involve only the corresponding simple properties of the fluctuating motion, like the extra rates of transfer of heat or momentum produced by the turbulence. There is thus a strong temptation to pursue a sub-branch of turbulence studies, dedicated to predicting the mean flow without actually bothering much about the turbulence: this is undoubtedly lifemanship, and will not suffice except in rather simple flows. To improve on this pragmatic approach requires some physical understanding of turbulence, to give one a feeling for the best assumptions to make and the best measurements to do.

Even our present incomplete understanding of the physics of turbulence is not easily accessible to the engineer and the experimental research worker, because there seems to be a gap between general books on fluid dynamics, which for reasons of space are forced to adopt the "lifemanship" approach, and specialist books on turbulence, which are conceptually and mathematically hard going. The measurement of turbulence is also hard going for the beginner, although the standards of patience and endurance necessary have decreased markedly in the last few years with the appearance of reliable electronic apparatus and innovations in the design of sensing elements. Since measurement and understanding go hand in hand, and since the simplest way of mastering the statistical analysis of turbulence is to ask how the measuring instruments analyse the output from an anemometer, I have tried in this book to combine an introduction to the physics of turbulence with a discussion of measurement techniques.

I have not found it necessary to use any heavy mathematics—the worst being a very elementary discussion of the relation between spectra and correlations—and I hope that the book will be understandable by anybody with a general background in engineering fluid dynamics. It covers all the ground likely to be touched in undergraduate courses and should be a useful introductory text for post-graduate courses. The first part, at least, may be useful to mathematicians or research workers who will go on to the specialist books mentioned above or to branches of the subject, like aeronautical acoustics, meteorology or pollution diffusion, which I have been able to cover only briefly. Most of all, I have tried to cater for the experimentalist, in college or industry, who has been assigned a turbulence project and who asks himself "What should I measure and how should I measure it?"

Turbulence and its measurement are both controversial subjects: many people take a more pragmatic view of turbulence and a less pragmatic view of measurement techniques than I do. I hope that I have made it clear when I have expressed my own point of view, and I hope that I have given the reader enough information for him to make up his mind for himself.

In reprinting, a few minor corrections and clarifications have been made and some key references added.

Acknowledgements

A FIRST draft of this book was used as lecture notes at a course on hot-wire anemometry run by Disa Elektronik A/S at the University of Surrey in 1969. I am grateful to Messrs. A. L. Cussens and D. H. S. Jones of Disa U.K. for their assistance, but in acknowledging their helpful criticisms of the draft I do not seek to implicate them or their company in the final result.

I am grateful for comments and criticisms from Prof. J. H. Horlock and his students, Mr. R. F. Johnson, Prof. D. M. McEligot, Mr. B. Slingsby and several of my students.

I am grateful to Mr. J. O'Leary for taking the frontispiece photographs, and to Sheila Bradshaw for help with the manuscript and references.

The quotations at the beginning of Chapters 1 to 8 are from *The Hunting of the Snark: an Agony in Eight Fits*, by Lewis Carroll.

Glossary

THIS list contains words used in this and other publications on turbulence and electronics and not defined in the main text: the definitions are mostly loose and are intended only as an aid to reading.

1. Turbulence

Average. An average with respect to *time* is defined mathematically as

$$\bar{f} = \operatorname*{Limit}_{T \to \infty} \frac{1}{2T} \int\limits_{-T+t}^{T+t} f(t') \, dt'.$$

A "statistically stationary" flow is one in which such an average is independent of t, the time at the mid-point of the averaging period. If $\bar{f}$ does depend on t (non-stationary flow) time averages are not useful and we resort to an ensemble average (see the end of Section 1.4).

Central limit theorem. See Section 2.1.

Correlation coefficient. The coefficient of correlation between two functions of time with zero averages u and v say, is $\overline{uv}/\sqrt{(\overline{u^2 v^2})}$: it lies between -1 and 1.

Covariance. The covariance of u and v is the mean product $\overline{uv}$.

Diffusivity. The ratio of a flux, or transport rate, to the mean gradient of the quantity being transported. If the transport is molecular, the diffusivity is a property of the fluid; if turbulent, the diffusivity obeys no simple laws and is generally different for transport in different directions. Diffusivity based on transport by mean velocity only has no significance.

Eddy. This is a designedly vague term: it is used to describe a typical motion in turbulent flow, usually covering a range of wavelengths of less than, say, 10 to 1, so that we can talk of "large" and "small" eddies in the same volume of fluid.

Ensemble. A set of samples of a random process (e.g. $u(x, y, z)$ for a set of different values of t or a set of repeats of an experiment). An *ensemble average* is an average over repeats of an experiment: we would *have* to use ensemble averages when studying, say, the turbulent boundary layer on a rapidly oscillating body. Ensemble averages are often used in mathematical discussions but the *ergodic hypothesis* says, roughly speaking, that ensemble averages and time averages are equivalent if the process is statistically stationary.

Flux. A "flow" of some quantity other than mass.

Gaussian. A Gaussian or "normal" process is one with the probability distribution

$$P(u) = \frac{1}{\sqrt{(2\pi\overline{u^2})}} \exp\left(-\frac{1}{2}\frac{u^2}{\overline{u^2}}\right)$$

where $\overline{u^2}$ is the variance. The central limit theorem ensures that most random molecular phenomena are Gaussian. Turbulence is significantly non-Gaussian, consisting of a large number of loosely related processes (eddies).

Homogeneous. Statistically independent of position in space, e.g. independent of x. Not necessarily isotropic.

Irrotational. Having zero vorticity.

Isotropic. Statistically independent of direction e.g. $\overline{u^2} = \overline{v^2} = \overline{w^2}$. In practice isotropic turbulence must also be homogeneous.

Local isotropy. Isotropy of the small-scale eddies.

Mean. Average (time or ensemble): *not*, at least in this book, a spatial average unless so stated.

Normal. (1) Normal or direct stress (taken positive if a tension). (2) Normal process: equals Gaussian process.

Probability. If $P(u')\,du$ is the fraction of total observation time that a function of time, u, spends between values u' and $u' + du$, the function $P(u)$ is called the probability distribution of u.

Process. Any sequence of events (usually restricted to a sequence that has to be described by statistical quantities) including continuous sequences like $u(t)$.

Random. Not simply (e.g. periodically) dependent on the independent variables, but *not* necessarily discontinuous.

Realization. One complete sample of a random process (see **Ensemble**).

Reynolds number. Ratio of typical acceleration to typical viscous stress gradients: if typical length is L and typical velocity U, Reynolds number is $(U^2/L)/(\nu U/L^2)$ $= UL/\nu$. Many different Reynolds numbers can be defined for turbulent flow.

Scale. A typical value of a dimension (usually velocity or length): for instance, the length scale of the largest eddies in a pipe flow is the pipe radius or diameter.

Standard deviation. Square root of variance.

Stationarity. The state of being stationary.

Stationary. *Statistically* independent of time (or of one or more space coordinates), e.g.

$$\mathop{\mathrm{Lt}}_{T\to\infty} \int_{T_0}^{T_0+T} u^2(t)\,dt$$

independent of T_0, so that $\overline{u^2}$ is independent of T_0.

Statistical. Relating to probabilities or average properties, e.g. $\overline{u^2}$, rather than randomly fluctuating properties, e.g. $u(t)$.

Stochastic. (Greek *stochos*, guess). Relating to random processes.

Structure. The dimensionless properties of a turbulence field, e.g. correlation shapes or ratios like $\overline{v^2}/\overline{u^2}$, as opposed to absolute values like integral scales or intensities.

Turbulence. See p. 17.

Variance. Mean square (see **Gaussian**).

Vortex line. A vortex of infinitesimal diameter, not necessarily straight and *not* necessarily having an angular velocity discontinuity between it and the surrounding fluid (Fig. 7).

Vortex sheet. A surface made up of many adjacent, non-intersecting vortex lines.

2. *Electronics*

Attenuator. Any network, usually of resistances only, whose output voltage or current is a fixed or adjustable fraction of the input voltage or current. "Attenuation" is defined as for amplification but is *less* than unity.

Audio frequency. An audible—as opposed to radio—frequency; roughly, less than 20 kHz.

Bandwidth. The frequency range between "cutoff points", i.e. the range within which an amplifier, filter or other circuit will pass an electrical signal; usually defined by the "upper and lower -3 dB points", the frequencies at which the signal falls to $1/\sqrt{2}$ of its ideal amplitude.

Cutoff. See **Bandwidth**.

Decibel. A "bel" (named after Alexander Graham Bell) is a logarithmic measure of electrical power ratio, the ratio in "bels" being the logarithm, to base 10, of the numerical ratio: a decibel is one-tenth of a bel, that is the power ratio in *decibels* (dB) is $10 \log_{10}$ (numerical ratio). Since the power dissipated in a given resistor is proportional to (voltage)^2, the custom has arisen of referring to a *voltage* ratio in dB as $20 \log_{10}$ (numerical voltage ratio), sometimes as "dB re 1 volt" or other reference but more usually as the ratio of two arbitrary voltages: for instance, a "gain of 40 dB" is an amplification (output voltage/input voltage) of $100:1$, and a "frequency response flat to within ± 3 dB" is a voltage amplification (or other transfer function) that varies by less than a factor of $\sqrt{2}$ (since $\log_{10} 2 \simeq 0.3$).

Feedback. An arrangement in which part of the output voltage of a circuit is added to the input voltage (positive feedback) or subtracted from it (negative feedback). The most frequent use in in constructing amplifiers whose gain is determined by fixed resistors and not by the characteristics of the transistors used: for instance, the gain of the circuit of Fig. 39 with $e_2 = 0$ is $e_0/e_1 \simeq -R_3/R_1$, virtually independent of the gain of the actual amplifier (the triangle) as long as the latter is much larger than R_3/R_1. It is necessary to compromise between high stability and a small number of amplifier stages.

Filter. A network whose bandwidth is pre-set or adjustable. A "low pass" filter (Fig. 42) passes frequencies from zero, or very low frequency, up to some chosen "cutoff" point; a high pass filter passes frequencies from some chosen cutoff point to a very high frequency.

Gain. Amplification of voltage or current, de_{out}/de_{in} or di_{out}/di_{in}.

Hertz (Hz). The preferred unit of frequency, cycles per second.

Octave. A range of $2:1$ in frequency (eight natural notes on the musical scale). The asymptotic rate at which a filter cuts off is often quoted as "dB per octave". Calculation reveals that "6 dB/octave" means "voltage $\propto 1/\text{frequency}$".

Phase shift. The phase angle between the voltage and current of a sinusoidal signal, or between two voltages, especially the input and output voltages of an amplifier.

Signal. Any voltage or current applied to a circuit.

Transconductance. The "gain" of a power amplifier, di_{out}/de_{in}. The unit is sometimes called the "mho", for $(\text{ohm})^{-1}$.

CHAPTER 1

The Physics of Turbulence

"Just the place for a Snark!" the Bellman cried,
    As he landed his crew with care;
Supporting each man on the top of the tide
    By a finger entwined in his hair.

"Just the place for a Snark! I have said it twice:
    That alone should encourage the crew.
Just the place for a Snark! I have said it thrice:
    What I tell you three times is true."

THE definition of turbulence will be found on p. 17, as the conclusion of the discussion which follows.

1.1. "Control-volume" Analysis for the Equations of Motion

By far the simplest way of deriving and explaining the equations of motion of a fluid is to consider the fluid in, or passing through, an imaginary fixed infinitesimal "control volume" with sides of length dx, dy and dz in the three coordinate directions (Fig. 1). We can pass by very simple steps from physical principles, like conservation of mass, momentum and energy, to the corresponding mathematical equations, which look extremely forbidding when written down without explanation. The main reason for the complication of the equations is that in discussions of turbulence we usually have to consider their most general form in which everything appears in triplicate. For instance, we have to deal with three momentum-conservation equations, one for each component, each of which will depend on three coordinate directions. Mathematicians

1

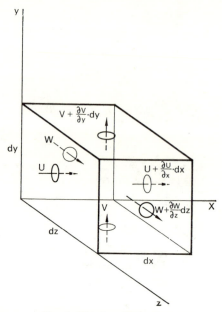

FIG. 1. Fixed control volume.

have their own shorthand, tensor notation, to make one term in an equation stand for three, or even for nine, but newcomers to a subject are often confused by having to learn a new notation as well as new physics. Therefore in this book I have usually written down only one term, or one equation, in a group of three, using the familiar x, y, z notation, and left the reader to infer the other terms by changing the variables in "cyclic order" ($x \to y$, $y \to z$, $z \to x$ and similarly for U, V and W): this is a watered-down tensor notation, more suitable for studying flows with a preferred direction and more likely to teach the student what the terms really mean.

As an example of control-volume analysis for those who are unfamiliar with it, we consider the law of conservation of mass, choosing the simple case of a fluid of constant density far from any free surface, so that the mass of fluid in the control volume is constant. We want to write down the condition that the fluid shall leave the control volume at the same rate that it enters. Now although fluid that enters through a given face may or may not leave the volume through the face that is parallel with the first, it is simplest to consider the difference in rate of flow through a pair of parallel faces and then to add up the differences for the three pairs to give a sum which we equate to zero. The rate at which mass flows through the left-hand face with sides dy, dz is $\varrho U \, dy \, dz$, with dimensions of mass per unit time, where U is the velocity in the x direction at the mid-point of the face. The rate of mass flow through the face parallel to this is $\varrho(U + (\partial U/\partial x) \, dx) \, dy \, dz$ since the U-component velocity increases by $(\partial U/\partial x) \, dx$ between the faces, so that the net outflow in the x direction is $\varrho(\partial U/\partial x) \, dx \, dy \, dz$ and the total net outflow in all three directions is

$$\varrho \left(\frac{\partial U}{\partial x} + \frac{\partial V}{\partial y} + \frac{\partial W}{\partial z} \right) dx \, dy \, dz.$$

The equation

$$\frac{\partial U}{\partial x} + \frac{\partial V}{\partial y} + \frac{\partial W}{\partial z} = 0$$

is the mass-conservation or continuity equation for an incompressible fluid: it is valid whether the flow is changing with time or not. In vector notation (which, incidentally, we shall not use very much in this book) it is written div $\mathbf{V} = 0$ where div is the divergence operator, so called from its physical significance in this and other equations, and $\mathbf{V}$ is the velocity vector. Readers who spotted that we ignored the variations of U in the y and z directions when calculating the flow through the face $dy \, dz$ should verify for themselves that $\partial U/\partial y$ and $\partial U/\partial z$ disappear when we average over the face $dy \, dz$ and that, according to the usual principles of the infinitesimal calculus, we can make dx, dy and dz small enough for second derivatives to be unimportant.

Another simple use of control volume analysis is to calculate the acceleration of the fluid as it passes through the control volume. The increase of U-component velocity between the two faces $dy \, dz$ is just $(\partial U/\partial x) \, dx$ (see Fig. 1 again) and since it takes a time dx/U for the fluid to cross between the two faces the consequent component of acceleration in the x direction is $U(\partial U/\partial x)$. Now if the fluid also has a V component of velocity, U will increase by $(\partial U/\partial y) \, dy$ in the time taken for the fluid to cross the control volume in the y direction, giving another contribution to the acceleration *in the x direction* $V(\partial U/\partial y)$: adding the third term $W(\partial U/\partial z)$ and remembering that in general $\partial U/\partial t$ at a fixed point may be non-zero, we find that the component in the x direction of the acceleration following the motion of the fluid is

$$\frac{\partial U}{\partial t} + U\frac{\partial U}{\partial x} + V\frac{\partial U}{\partial y} + W\frac{\partial U}{\partial z}$$

and similarly for the y and z components of the acceleration.

1.2. Newton's Second Law of Motion

Fluids, like solids, obey Newton's second law "force = rate of change of momentum" which is the conservation law for momentum: note that momentum is a vector quantity, having components in the x, y and z directions. The rate of change of momentum of the fluid in the control volume, that is, the product of its mass and its acceleration as derived above, is equal to the sum of the forces acting on the fluid. The forces are:

(1) Externally applied "body forces" like magnetic fields or gravity which act uniformly on the fluid in the control volume: for instance, the force due to gravity is $\varrho g\, dx\, dy\, dz$.

(2) Pressure gradients, which can be regarded as producing a net force on the fluid between any two parallel faces of the control volume [e.g. $(-(\partial p/\partial y)\, dy)\, dx\, dz$ between the faces perpendicular to the y-axis in Fig. 1]. Often, the only effect of a body force is to produce an opposing "hydrostatic pressure gradient", for example $\partial p/\partial y = -\varrho g$ in a gravitational field with g in the negative y direction, so that both can be neglected. However pressure gradients can be generated by the motion of the fluid so that the pressure p must be regarded as an independent variable, just like the three velocity components U, V and W in the x, y and z directions.

(3) Stress gradients caused by deformation of the fluid. A "fluid" is defined as a substance in which stresses, other than the hydrostatic pressure, depend on rate of deformation ("rate of strain") rather than on strain itself as in a solid: there are some borderline substances in which stresses depend on both strain and rate of strain, and even on the details of past strains. We use the notation σ_{xy} for the component of stress in the x direction acting on a surface perpendicular to the y direction, and similarly for other stresses.

Before we can write down the equations of motion we must know, or guess, the relation between stress and rate of strain for the fluid.

1.3. The Newtonian Viscous Fluid[1]

The simplest possible relation between stress and rate of strain is direct proportionality, the factor of proportionality being the viscosity μ. In a Newtonian viscous fluid, the viscosity is a property of the fluid, depending only on the temperature and (very slightly) on the pressure. Virtually all gases and most liquids are closely Newtonian. If in Fig. 2 the velocity in the x direction, U, has a gradient $\partial U/\partial y$ in the y direction, other rates of strain being negligible (Fig. 2b) then a shear stress $\sigma_{xy} = \mu(\partial U/\partial y)$ acts in the x direction on the two faces perpendicular

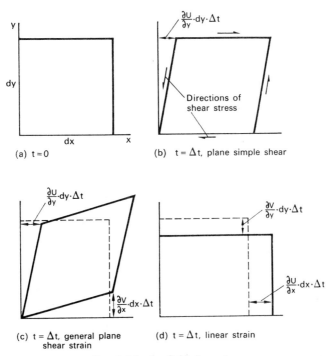

(a) t = 0

(b) t = Δt, plane simple shear

(c) t = Δt, general plane shear strain

(d) t = Δt, linear strain

FIG. 2. Moving fluid element.

to the y direction and an equal shear stress $\sigma_{yx} = \sigma_{xy}$ acts on the other two faces as shown. This is a "plane simple shear", $(1/2)\,\partial U/\partial y$ being the rate of (shear) strain e_{xy}. Note that the shear stresses can be resolved into a normal (tensile) stress along the longer diagonal of the infinitesimally distorted element of Fig. 2b, and a compressive stress along the shorter diagonal: these stresses are proportional to the rates of extension or compression of the respective diagonals, $\pm(1/2)\,\partial U/\partial y$. The fact that the diagonals are also rotating (clockwise in Fig. 2b) does not affect the stress. In Fig. 2c the rate of extension or compression of the two diagonals is found to be $\pm(1/2)\,(\partial U/\partial y + \partial V/\partial x)$ and the corresponding shear stress is $\mu(\partial U/\partial y + \partial V/\partial x)$. In general, therefore, $\sigma_{xy} = \sigma_{yx} = \mu(\partial U/\partial y + \partial V/\partial x)$ and similarly with x and y replaced by any other symbols. Now let us consider a simple linear strain (Fig. 2d). Using the above formula with y replaced by x we find that the normal (tensile) stress in the x direction, σ_{xx}, is $2\mu\,\partial U/\partial x$: similarly, the normal stress in the y direction, σ_{yy}, is $2\mu\,\partial V/\partial y$, which is equal to $-2\mu\,\partial U/\partial x$ if the strain is confined to the xy-plane so that $\partial U/\partial x + \partial V/\partial y = 0$ by continuity. The diligent reader will have noticed that the factor of proportionality in the relation between the normal stress and the rate of linear strain along the diagonal in Fig. 2b is also 2μ: we *define* the linear rates of strain in Fig. 2d as $e_{xx} = \partial U/\partial x$ and $e_{yy} = \partial V/\partial y$.

Any strain field can be resolved into linear strains along, and shear strains in planes perpendicular to, the three coordinate axes (together with components of rotation about the axes). Therefore the corresponding normal and shear stress components are given by

$$\sigma_{xy} = 2\mu e_{xy} = \mu(\partial U/\partial y + \partial V/\partial x) \quad \text{(shear stress in a plane perpendicular to the z-axis)}$$

and

$$\sigma_{xx} = 2\mu e_{xx} = \mu(\partial U/\partial x + \partial U/\partial x) = 2\mu\partial U/\partial x \quad \text{(tensile stress along the x-axis)}$$

and the analogous formulae for the other axes (we follow the definitions of Rosenhead[1]). Alternatively, the strain field can be resolved into three mutually perpendicular linear "principal" strains (and one component of rotation) not necessarily along the coordinate axes, with corresponding "principal" normal stresses.

In incompressible flow the sum of the normal stresses, $\sigma_{xx} + \sigma_{yy} + \sigma_{zz}$, is zero (using the continuity equation), so that viscosity does not affect the pressure p directly: this is more a matter of definition than a physical principle.

Constant stresses do not produce a net acceleration or rotation of the fluid: the net force on the fluid in the control volume is the difference between the forces on opposite faces, i.e. it depends on the stress *gradient*. The net force in the x direction in the case of a simple shear (Fig. 2b) is

$$\frac{\partial}{\partial y}\left(\mu \frac{\partial U}{\partial y}\right) dx\, dy\, dz, \quad \text{or} \quad \frac{1}{\varrho}\frac{\partial}{\partial y}\left(\mu \frac{\partial U}{\partial y}\right)$$

per unit mass. If the pressure and temperature changes in the fluid are large enough to change its viscosity and density appreciably, the expressions for the viscous stresses in the fluid become surprisingly complicated even for a simple Newtonian fluid. The kinetic theory of gases shows that just as stress (or momentum transfer) is proportional to rate of strain (or momentum gradient) so heat transfer is proportional to temperature gradient, and so on for other transport properties: the various factors of proportionality are called "diffusivities".

1.4. Possible Solutions of the Equations of Motion

The equations which express Newton's second law for a Newtonian viscous fluid are the Navier–Stokes equations. There are three, one for each of the components of momentum, and they are given for reference in Appendix 1. They can be derived by control volume analysis, most of the required information being given in the last section. The fourth equation, which completes the set for the four variables U, V, W and p, is the continuity equation derived in Section 1.1. An equation for the pressure in terms of U, V and W can be deduced by rearranging the equations and is also given in Appendix 1 (we shall not say much about the fluctuating pressure in this book: this is partly because it is determined by the velocity field—although the equation for the pressure is too difficult to solve in general—but also because there is no generally accepted technique of measuring the pressure fluctuation within a turbulent flow—although some progress has been made recently).

Despite their physical simplicity, the equations look—and are—very difficult to solve: the four variables U, V, W and p are in general functions of x, y, z and t, and the spatial scales of their variation can, in principle, take any values between those imposed by the size of the container in which the fluid is flowing and those of the molecular motion. Except in low-density gas flows and shock waves, viscosity prevents the spatial scales from extending down into the molecular range and the fluid can be regarded as a continuum: this implies that the above expressions for viscous stress are valid even if the flow is varying rapidly in space and time. The most general case of fluid motion, exhibiting all the complications allowed by the Navier–Stokes equations, is called turbulence: complete solution is far beyond the capacity of present-day computers, so that the study of turbulence leans heavily on experiment. Most of the flows found in nature or in engineering are turbulent, so that the study of turbulence is one of the most important branches of fluid mechanics.

Occasionally one can find, mathematically or experimentally, flows that are independent of time, or very simply dependent on time, which means that the spatial dependence is also comparatively simple: we call these laminar flows. Even rarer, but important, are "inviscid" flows in

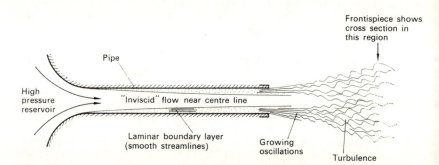

FIG. 3. Flow through short pipe at moderate Reynolds number (at high Reynolds number the flow would go turbulent within the pipe).

which viscous stresses are negligible. Even a steady laminar or "inviscid" flow may become unstable to small disturbances (Fig. 3) so that growing oscillations of velocity about the mean, either in space or in time, or both, occur (on the elementary level of this book, one cannot learn much about turbulence from studying the instability of steady flows, so we shall not treat the subject further). As the oscillations grow, they gradually change from a simple sinusoidal form to a final, apparently random, eddying motion covering a continuous range of wavelengths and frequencies.

This *randomness* is the essential characteristic of turbulence. Of course, the velocity is still a continuous function of space and time, and statistical correlations between the motions at different points can be distinguished[2], but the probability distribution of the velocity at a given point is nearly the familiar Gaussian or normal probability distribution that occurs in electronic noise and other random processes. This means that we have to treat turbulence in terms of statistical properties because we cannot cope, either mathematically or experimentally, with the information needed in a full treatment. The simplest statistical property is the average, with respect to time, of the velocity at a point, called the mean velocity. If we could measure only the mean velocity, we should see no qualitative difference between laminar and turbulent flows. The fluctuations of the velocity about the mean value are typically fairly small, say up to ± 10 per cent of the mean velocity near the axis of a pipe: however, the pressure drop down the pipe for a given flow rate may be ten or a hundred times larger than if the flow could be persuaded to remain laminar.

One (uncommon) case that sometimes causes difficulty to students is exemplified by turbulent flow through a pipe from a reservoir whose pressure is steadily decreasing in time (Fig. 4). How do we define mean velocities? A time average is not very meaningful because we have obviously to treat the long-term trend and the short-term fluctuations separately. All we can do is to repeat the experiment many times, each time measuring the instantaneous velocity at a point at the instant the supply pressure passes—say—50 cm water gauge: the average of all these instantaneous measurements (the so-called ensemble average) is the appropriate mean velocity at the point for a supply pressure of 50 cm water. This type of flow is called non-stationary, to distinguish it from the "statistically stationary" flows (constant supply pressure) for which simple time averages can be defined. In the rest of the book we shall talk about time averages, leaving the reader whose concern is with non-stationary flows to substitute the words "ensemble averages".

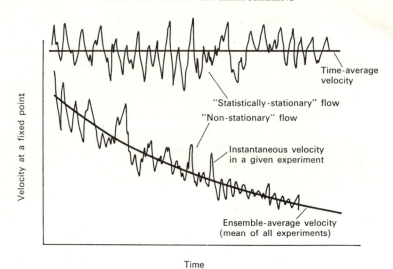

FIG. 4. Comparison of ordinary and time-dependent turbulent flow (e.g. pipe flow with steady or decreasing supply pressure).

1.5. The Reynolds Stresses

We now come to the explanation of how quite small velocity fluctuations can produce such large changes in flow resistance and other properties. The rate at which x-component momentum passes through one of the faces $dy\,dz$ of the elementary control volume of Fig. 1 is equal to the product of the mass flow rate $\varrho U\,dy\,dz$ and the velocity U: we call this the momentum flux through the face. Now suppose that in addition to a mean velocity U there is a time-dependent (fluctuating) component of velocity in the x direction, u—the mean (time average) value of u, denoted by $\bar{u}$, being zero by definition. Then the x-component momentum flux through the face $dy\,dz$ is

$$\varrho(U + u)^2\,dy\,dz \quad \text{or} \quad \varrho(U^2 + 2Uu + u^2)\,dy\,dz$$

equal in the mean to $\varrho(U^2 + \overline{u^2})\,dy\,dz$ where $\overline{u^2}$ is the mean value of u^2. Therefore, a fluctuation with zero mean, superimposed on the mean velocity, produces a mean momentum flux of its own, proportional to the mean square of the fluctuating velocity, because momentum flux is the product of mass flow and velocity and the fluctuation contributes to both. This non-linearity of the relation between velocity and momentum flux appears in the Navier–Stokes equations and is the basic cause of their mathematical difficulty even in laminar flow.

If we measure the mean velocity and pressure in turbulent flow we find that the mean motion is not that which would be produced by viscous forces alone: there is an extra apparent stress $-\varrho\overline{u^2}$ normal to the face $dy\,dz$ (it is a compressive stress since $\overline{u^2}$ is positive: in the notation of Section 1.3 it is an addition to σ_{xx}). The equivalence of stress and momentum flux follows at once from Newton's second law. Similarly, there are extra normal stresses $-\varrho\overline{v^2}$ and $-\varrho\overline{w^2}$ in the y and z directions: usually, $\overline{u^2}$, $\overline{v^2}$ and $\overline{w^2}$ differ by no more than a factor of 2 or 3.

Again, the rate at which x-component momentum passes through the face $dx\,dz$ of the control volume is the product of the mass flow in the y direction, $\varrho(V + v)\,dx\,dz$, and the velocity in the x direction, $U + u$: the mean momentum flux is $\varrho(UV + \overline{uv})\,dx\,dz$ and $\sigma_{xy} = -\varrho\overline{uv}$ represents an extra mean shear stress on the face $dx\,dz$. Note that the rate at which y-component momentum passes through the face $dy\,dz$ leads to a shear stress $\sigma_{yx} = \sigma_{xy}$ on that face also, just as in the case of viscous stresses (which can, of course, exist at the same time as the extra turbulent stresses). These extra turbulent stresses are called the Reynolds stresses, in honour of Osborne Reynolds: the time-mean Navier–Stokes equations, in which these stresses appear, are called the Reynolds equations.

Turbulence produces additional fluxes (rates of transport) of quantities other than momentum: for example, if there are temperature fluctuations, θ, in the flow, implying enthalpy fluctuations $\varrho c_p\theta$ per unit volume (assuming ϱ and c_p to be constant for simplicity) then there is an extra rate of enthalpy transfer $\varrho c_p\overline{\theta u}$ in the x direction, $\varrho c_p\overline{\theta v}$ in the y direction and $\varrho c_p\overline{\theta w}$ in the z direction, in addition to enthalpy transfer by molecular conduction.

Now velocity fluctuations of ± 10 per cent about the mean, U, will produce Reynolds stresses of the order of $0 \cdot 001\varrho U^2$. In a pipe, say, mean velocity gradients are of order U/d, giving viscous stresses of order $\mu U/d$. The ratio is $0 \cdot 001\,Ud/\nu$, or 100 if $Ud/\nu = 100{,}000$. Indeed, in nearly all cases of interest, Reynolds stresses and other turbulent transport rates are much larger than viscous stresses and other molecular

transport rates: this is why turbulence is of such great practical importance as well as being a fascinating phenomenon. At first sight, the
mechanism by which turbulent eddies produce the Reynolds stresses
seems quite similar to the mechanism by which random molecular
motion produces viscous stresses, but there is no worth-while analogy
between the two because

(1) turbulent eddies are continuous and contiguous whereas gas
molecules are discrete and collide only at intervals;
(2) although molecular mean free paths are small compared to the
dimensions of the mean flow, turbulent eddies are not.

It follows that the turbulent transport rates are *not* determined by the
mean gradient of the transported quantity as in molecular transport—
that is to say, the turbulent diffusivities, defined analogously to the
molecular diffusivities mentioned at the end of Section 1.3, are *not*
usually constants or even discoverable functions of the local variables,
but depend on the previous history of the flow which carries the turbulent eddies (see Section 2.2). There are some special cases in which
the diffusivities depend simply on the velocity and length scales of the
flow when the latter are particularly simple (*not* because the eddy motion
is any simpler than usual): we shall return to this subject in Sections 3.3
and 3.4.

Since the extra apparent stresses and transport rates are produced
by the fluctuating motion itself, it appears that in order to understand
them we must at least partly understand the behaviour of the fluctuations. Because we cannot solve the complete time-dependent Navier–
Stokes equations for each case we have to appeal to experiment even for
a qualitative understanding of turbulence, as well as for quantitative
information for engineering purposes. However, it is possible to do
theoretical work on turbulence at many levels, ranging from attempts
to solve the Navier–Stokes equations in simplified form to elementary
dimensional analysis. A very brief account of the present state of this
work, and an introduction to current trends in experimental research,
is given in Appendix 2.

1.6. Vortex Stretching[3]

Turbulent eddies (and some laminar flows or "inviscid" flows) have
both translational and rotational motion, familiar to anyone who has

looked over a river bridge. The net rate of rotation (or average angular velocity) about the z axis of the fluid element shown in Fig. 2c is

$$\frac{1}{2}\left(\frac{\partial v}{\partial x} - \frac{\partial u}{\partial y}\right)$$

(we use small letters to show that in general we are dealing with fluctuations). We define the z component of *vorticity* as twice this angular velocity, $(\partial v/\partial x) - (\partial u/\partial y)$. Distinguish the vorticity from (twice) rate of shear strain $(\partial u/\partial y) + (\partial v/\partial x)$: one is a measure of rotation, the other a measure of deformation (Fig. 5). Now if in addition to a rotation about the z-axis the fluid element is under the influence of a rate of linear strain in the z direction, $\partial w/\partial z$, the element will be stretched in the z direction and its cross-section in the xy-plane will get smaller. If

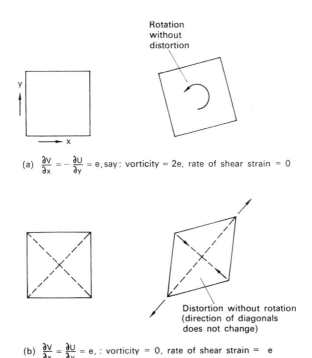

Rotation without distortion

(a) $\dfrac{\partial v}{\partial x} = -\dfrac{\partial u}{\partial y} = e$, say: vorticity $= 2e$, rate of shear strain $= 0$

Distortion without rotation (direction of diagonals does not change)

(b) $\dfrac{\partial v}{\partial x} = \dfrac{\partial u}{\partial y} = e$, : vorticity $= 0$, rate of shear strain $= e$

FIG. 5. The distinction between vorticity and rate of (shear) strain.

we take the case of an element of circular cross-section in the xy-plane and neglect viscous forces for simplicity, we can see that conservation of angular momentum requires the product of the vorticity and the square of the radius to remain constant: more generally, the integral of the tangential component of velocity round the perimeter, called the circulation, remains constant in the absence of viscous forces (see pp. 93 ff. of Batchelor[1]). During the stretching process the kinetic energy of rotation increases (at the expense of the kinetic energy of the w component motion that does the stretching) and the scale of the motion in the xy-plane decreases. Therefore an extension in one direction (the z direction here) can decrease the length scales and increase the velocity components in the other two directions (x and y) which in turn stretch other elements of fluid with vorticity components in these directions, and so on. The length scale of the motion that is augmented gets smaller at each stage. If we draw out a family tree (Fig. 6) showing how stretching in the z direction intensifies the motion in the x and y directions, producing smaller-scale stretching in x and y and intensifying the motion in the y, z and z, x directions respectively, and so on, we can see qualitatively

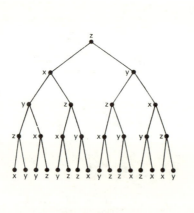

Frequency of symbols
at each generation

x	y	z
0	0	1
1	1	0
1	1	2
3	3	2
5	5	6
11	11	10
21	21	22

FIG. 6. "Family tree" showing how vortex stretching produces small-scale isotropy. The labels are the directions of stretching in each "generation": the length scale decreases from one generation to the next.

that an initial stretching in one direction produces nearly equal amounts of (smaller-scale) stretching in each of the x, y and z directions after a few stages of the process. Thus the small-scale eddies in turbulence do not share the preferred orientation of the mean rate of strain: they have a near universal structure which makes their study easier (see Section 2.5). The "cascade" of energy of the turbulent motion continues to smaller and smaller scales (larger and larger velocity gradients): indeed, discontinuities of velocity would develop if it were not for the smoothing action of viscosity. Put another way, viscosity finally dissipates (into thermal internal energy, loosely called "heat") the energy that is transferred to the smallest eddies, but it does not play any essential part in the stretching process as such.

In the above discussion it was implied that the element of fluid considered was part of a line vortex with its axis in the z direction (see Fig. 7 for a brief revision of vortex properties). We can imagine any flow with vorticity to be made up of large numbers of infinitesimal slender vortices, "vortex lines": sometimes it is convenient to talk of

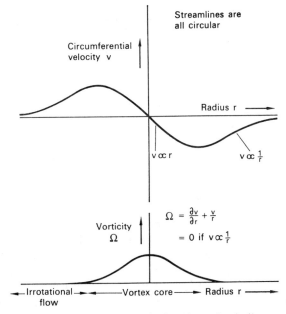

FIG. 7. Velocity and vorticity distributions in a simple line vortex.

a "vortex sheet", which is a layer of locally parallel vortex lines—for instance, laminar shear layers are often discussed as if they were made up of a stack of elementary vortex sheets, with the vortex lines parallel or nearly so.

Turbulence can be thought of as a tangle of vortex lines or partly rolled vortex sheets, stretched in a preferred direction by the mean flow (the mean vortex lines) and in random directions by each other. Turbulence *always* has all three directions of motion even if the *mean* velocity has only one or two components: if the fluctuating velocity component in one direction were everywhere zero the vortex lines would necessarily all lie in this direction and there would be no vortex stretching, no transfer of fluctuation energy to smaller scales, and the motion would not be what we call turbulence. But for the diffusing effect of viscosity, vortex lines or sheets would move with the fluid: the effect of viscous diffusion is seen in the slow growth of laminar shear layers. In turbulent flow, viscous diffusion of vorticity is negligible except for the smallest eddies—those that dissipate the energy transferred from the larger eddies. Fluid that is initially without vorticity ("irrotational") can acquire it only by viscous diffusion but, once acquired, vorticity can be increased many orders of magnitude by vortex stretching. Pressure fluctuations do *not* directly affect vorticity in incompressible flow.

The rate of supply of kinetic energy to the turbulence is, in the absence of body forces, the rate at which work is done by the mean rate of strain against the Reynolds stresses in the flow as it stretches the turbulent vortex lines. In laminar flow, viscous stresses caused by molecular motion convert ("dissipate") mean flow kinetic energy directly into thermal internal energy: in turbulent flow the eddies extract energy from the mean flow and retain it for a while before it reaches the small dissipating eddies. Turbulent kinetic energy, $\frac{1}{2}\varrho(\overline{u^2} + \overline{v^2} + \overline{w^2})$ per unit volume, is introduced into the eddies that contribute to the Reynolds stresses in rough proportion to those contributions. The stress-producing eddies are the larger ones, which are best able to interact with the mean flow: we have already seen that vortex stretching tends to make the smaller eddies lose all sense of direction and become statistically isotropic (see Glossary) so that, for instance, their contribution to the Reynolds shear stress $-\varrho\overline{uv}$ is zero. The smaller eddies are much weaker than those that produce most of the Reynolds stress because most of the energy that reaches them is immediately passed on to the smallest eddies of all and there dissipated by viscosity. In the central part of a typical pipe flow at least half the turbulent kinetic energy and most of the Reynolds shear stress occurs in eddies with wavelengths greater than the pipe radius. The size of the dissipating eddies depends on the viscosity and the speed of flow as well: typically their wavelength is less than 1 per cent of the pipe radius (see the research paper by Laufer in Section 3.6 of the "Further Reading" list for more details).

Since the viscous stresses are usually so small compared with the turbulent stresses and since the parts of the eddy structure that depend on viscosity are so small and so weak compared with the stress-producing part of the turbulence, we can for many purposes neglect viscosity in the study of turbulent flow, regarding it only as a property of the fluid that produces energy dissipation in very small eddies. Exceptions are flows in process of transition from laminar to turbulent and the flow very close to a solid surface (say within 0·5 mm in air flow in a pipe at 20 metres/sec). In both these cases viscous and turbulent stresses can be of the same order and viscosity directly affects the eddies that produce the Reynolds stresses. This makes transition a very difficult problem but fortunately the flow in the viscous sub-layer close to a solid surface is a function of only a few variables, and dimensional analysis, plus a few empirical constants, "solves" the problem for engineering purposes. Apart from these cases, the behaviour of turbulence would be much the same whatever the dissipation mechanism in the fluid. The main characteristics of turbulence are the result of three-dimensional vortex stretching which, mathematically speaking, depends on the non-linear terms in the Navier–Stokes equations that represent the acceleration of the fluid, and not upon the terms representing the viscous forces.

We can now define *turbulence*:

Turbulence is a three-dimensional time-dependent motion in which vortex stretching causes velocity fluctuations to spread to all wavelengths between a minimum determined by viscous forces and a maximum determined by the boundary conditions of the flow. It is the usual state of fluid motion except at low Reynolds numbers.

1.7. Compressible Flow

Another simplification in the study of turbulence is that its general behaviour is apparently unaffected by compressibility if the pressure fluctuations within the turbulence are small compared with the absolute pressure, that is, if the fluctuating Mach number, u/(speed of sound) say, is small. Since the velocity fluctuations are usually a small percentage of the mean velocity, this is the case for all mean-stream Mach numbers less than about 5. For the same reason, density fluctuations arising from temperature differences in a gas flow are small enough to have no direct effect on the turbulence if the change in absolute temperature across the flow is less than a factor of 5 or 6. Of course, mean density changes certainly affect the mean flow and may also have some effect on the turbulence, but the latter effects are small unless strong streamwise density gradients occur (because of large pressure gradients or chemical reactions). For these reasons we shall ignore compressibility in the main part of the book: unfortunately, the process of making *measurements* is made much more difficult by the presence of temperature or density fluctuations, and these difficulties will be discussed in Chapter 7, with reference to high-speed and low-speed flows.

1.8. Flow-visualization Experiments

Still photographs of smoke or dye traces in turbulent flow do not mean much to the inexperienced: ciné films are more helpful, and a recent one called "Turbulence", produced by Prof. R. W. Stewart of the University of British Columbia and distributed by Encyclopaedia Britannica Films Inc. for the (American) National Committee for Fluid Motion Films, is an excellent introduction to the subject. I urge the reader of this book to do his own flow visualization, starting by watching clouds, rivers and smoke plumes to see how turbulence diffuses matter and momentum and tangles a line of fluid coming from a fixed source: even watching smoke from a cigarette helps to pass time in dull committee meetings. Turbulence is such a common phenomenon that there is no shortage of natural examples: however, some simple experiments can be even more informative, even to people with a good theoretical grounding in the subject.

Fill a bucket from the water tap, and immediately release a thin stream of dye below the surface (a fountain pen will do). The dye will be drawn out into a tangled line whose total length vastly exceeds the distance the water has moved relative to the point of release[3]: if vorticity had been released instead of dye, it would have been increased many times by this stretching. After the turbulence caused by filling the bucket has died down, the dye filament will have spread to most parts of the bucket while still showing as a distinct line. Molecular diffusion scarcely thickens the line (although blobs of dye may tend to fall away if the dye is much heavier than water). Now stir vigorously for a second or two: in a few more seconds, the resulting intense turbulence will have mixed the dye completely with the water, something that would take hours or days for molecular diffusion to do unaided.

The behaviour of turbulent shear flow, discussed in Chapter 3, can be demonstrated more easily in air than in water. Make up the simple smoke generator shown in Fig. 8. Light the cigarette and blow down the tube (preferately with dry air): the resulting flow will look like that

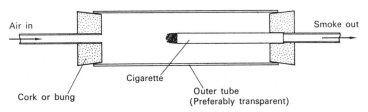

FIG. 8. Cigarette smoke generator.

shown in Fig. 3: the point at which transition from laminar to turbulent flow begins will move closer to the end of the tube the faster the blowing rate (i.e. the higher the Reynolds number). Blow the jet through a thin plane of light (use a slide projector with the slide blacked out except for a narrow slit) and notice how highly re-entrant the cross-section is (see Frontispiece). Try illuminating a plane parallel to the axis of the jet (this is more difficult to set up): tongues of smoke can be seen briefly extending from the main jet before being drawn back as the jet entrains air from the surroundings.

If the jet is blown tangential to a solid surface the turbulence close to the surface is flattened in side view: the velocity fluctuations and Reynolds stress are zero at the surface, so the surface shear stress is applied wholly by viscosity, leading to a very high velocity gradient near the surface. The same thing happens in boundary layers and pipe flows.

If a step is now put in the way of the wall jet the flow will separate ahead of the step, leaving a rather unsteady recirculating flow in the concave corner (Fig. 26). Turbulent flows can stand a much higher pressure rise before separating than laminar flows, so the recirculating flow region is shorter. Its unsteadiness would not show up in mean measurements. The flow round the end of the step is quite complicated, and near-steady trailing vortices can be seen: they come from the *mean* vorticity of the jet about an axis parallel to the step. This vorticity is stretched, and turned towards the direction of motion, as the flow rounds the corner. This mean vortex stretching accounts for the great complication of three-dimensional flows, whether laminar or turbulent.

Between them, these simple experiments demonstrate most of the qualitative features of turbulence described in this book. In particular, they should convince the reader of the need to study turbulence by statistical means, and of the essential implausibility of simple formulae purporting to explain this complicated phenomenon. They should also indicate that one cannot expect to find out all about turbulence from a few values of $\overline{u^2}$ (the quantity easiest to measure). However, one can make *some* sense of turbulence by combining theoretical and experimental approaches: the rest of the book is an introduction to both.

We may summarize this chapter by the definition on p. 17, together with the following properties of turbulence.

(1) Apparent mean stresses in turbulent flow are determined by the velocity fluctuations which, strictly speaking, depend on the whole history of the flow and not on the mean flow at the point considered.

(2) Viscous stresses are usually small compared with turbulent stresses: also, viscosity affects only the smallest eddies and the turbulent stresses are usually independent of viscosity.

(3) If the fluctuating part of the Mach number is much less than unity, turbulence is not much affected by compressibility unless large streamwise density gradients occur.

CHAPTER 2

Measurable Quantities and their Physical Significance

Come, listen, my men, while I tell you again
The five unmistakable marks
By which you may know, wheresoever you go,
The warranted genuine Snarks.

2.1. Statistics of Random Processes

The "central limit theorem"[2] states that the probability distribution of a continuous variable which is the sum of a large number of independent variables is approximately normal or Gaussian (Fig. 9: see Glossary

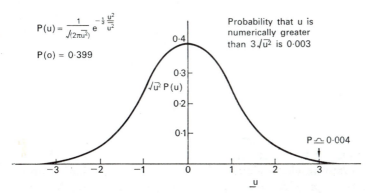

$$P(u) = \frac{1}{\sqrt{(2\pi u^2)}} e^{-\frac{1}{2}\frac{u^2}{u^2}}$$

$$P(o) = 0.399$$

$\sqrt{\overline{u^2}}\, P(u)$

Probability that u is numerically greater than $3\sqrt{\overline{u^2}}$ is 0.003

$P \simeq 0.004$

FIG. 9. Gaussian ("normal") probability distribution.

for definitions). A simple example is the clapping of a large audience; another is the noise of an amplifier, caused by random motion of electrons: the two sound exactly the same. Nearly all the random processes

21

that occur in nature or technology have Gaussian probability distributions or related ones like the Maxwell distribution, and the theory of Gaussian processes is highly developed. Unhappily, turbulence is an exception: certainly, it is the sum of a large number of processes (velocity fields of elementary vortex lines) but they are not quite independent—the audience is trying to give a slow handclap. The small departures from a Gaussian probability distribution are, moreover, the most interesting features of the turbulence: for instance, the triple products like $\overline{u^2 v}$, which would be zero in a Gaussian process, are connected with energy transfer by the turbulence, either from one point to another or one eddy size to another. Again, it is the non-linearity of the Navier–Stokes equations (p. 11) that is responsible.

Books on statistical mathematics and such linear random processes as electrical noise tend to concentrate on Gaussian theory, sometimes to the extent of presenting results that are valid only for Gaussian processes without making this restriction entirely clear. Also, elementary books usually start with discrete processes (successive events like the throwing of dice) which are even less relevant than continuous Gaussian processes to the statistics of turbulence. I have therefore tried to make this book self-contained rather than recommend a separate book on statistics. Definitions of the statistical terms will be found in the Glossary. Since readers without a mathematical background find a rigorous analysis very hard going I have generally given an illustration rather than a proof of the necessary statistical results. We are in all cases concerned with quantities that can be measured by fairly simple combinations of probes and electronic instruments and in all cases the easiest way of understanding the statistics is to ask "what do the instruments measure?" For instance, the time average of a fluctuating quantity is the reading of a well-damped voltmeter whose input is a voltage proportional to the quantity concerned, and the mean product of two fluctuating quantities is obtained by feeding two voltages into an electronic box whose output at any instant is proportional to their product, and then feeding that output to the well-damped voltmeter (actually one should use an electronic integrator rather than a well-damped meter but the two are closely equivalent, as shown in Section 6.4).

The statistical quantities that we usually want to measure in turbulence studies are those connected with

(i) the spatial distribution of the Reynolds stresses themselves, or of the turbulent kinetic energy per unit volume of the turbulence [which happens to be minus one-half the sum of the Reynolds normal stresses, $\frac{1}{2}\varrho(\overline{u^2} + \overline{v^2} + \overline{w^2})$]: it is helpful to think of the individual stresses as being "conserved", like energy;

(ii) the rates at which turbulent kinetic energy, or the individual Reynolds stresses, are produced, destroyed or transported from one point in space to another (Fig. 10 is a summary of the processes involved);

(iii) the contribution of different sizes of eddy to the Reynolds stresses;

(iv) the contribution of different sizes of eddy to the rates mentioned in (ii), *and* the rate at which energy or Reynolds stress is transferred from one range of eddy size to another.

The qualitative foundation for a discussion of these quantities was laid in Chapter 1, in the discussion of Reynolds stress (Section 1.5) and of the "cascade" of energy, from the larger eddies which are unaffected by viscosity to the smallest ones which are dominated by viscosity (Section 1.6). The material in Section 2.2, covering (i) and (ii) above, follows at once from this foundation. Unfortunately a quantitative discussion of (iii) and (iv) must be preceded by a careful definition of "eddy size" and this requires some mathematical analysis. The reader who has some knowledge of frequency spectra as used in electrical engineering, and who is prepared to accept the approximate definition "wave number = (radian frequency)/(mean velocity)", can pass straight to the discussion of wave-number spectra in Section 2.6.2: however, all readers should study Sections 2.3 to 2.6.1 eventually.

Many workers will not be concerned with the more complicated quantities mentioned in the latter part of the chapter (Sections 2.6–2.10) but it seemed worth while to collect in this chapter an expanded glossary of *all* the mathematical information that is needed for the analysis of experimental results: contrary to popular opinion, there is not very much of it. Theoretical papers on turbulence involve more complicated analysis but generally the authors are—or should be—discussing quantities with some physical meaning. Readers who want more detail should consult ref. 1. The best mathematical introduction is Chapters 2 and 3 of the book by Batchelor[2], who quotes references to the basic theorems that justify the statistical representations that we use.

2.2. Turbulent Energy

We have met $\overline{u^2}$, $\overline{v^2}$ and $\overline{w^2}$ as components of (Reynolds normal stress) $\div$ (density). They are also known as the mean-square components of turbulence *intensity*: often, measurements are presented as *root-mean-square* (r.m.s.) intensities, $\sqrt{\overline{u^2}}$, $\sqrt{\overline{v^2}}$, $\sqrt{\overline{w^2}}$, mainly because r.m.s. quantities are used in discussion of sinusoidal "signals" (voltages or currents) in electrical engineering, but the r.m.s. value is much less appropriate for random functions. The mean square of the sum of two fluctuating quantities u_1 and u_2 is $\overline{(u_1 + u_2)^2} = \overline{u_1^2} + \overline{u_2^2} + 2\overline{u_1 u_2}$; if the correlation coefficient $\overline{u_1 u_2}/\sqrt{(\overline{u_1^2 u_2^2})}$ is unity (for instance, if $u_1 = a_1 \sin \omega t$, $u_2 = a_2 \sin \omega t$) then the root mean square of the sum is the sum of the root-mean-squares, $(a_1 + a_2)/\sqrt{2}$, which is a useful property of the r.m.s.; however, if the correlation coefficient is not unity, no such simple relation holds, and the statistical description is most easily made in terms of the mean squares and mean product (it is always these quantities, and not *root* mean squares, that appear in equations with physical meaning).

It is worth noting that there is no fundamental difference between normal stress and shear stress: for instance, $(u + v)/\sqrt{2}$ is the component of the velocity fluctuation along a line in the xy-plane at 45 deg to the x-axis, so that its mean square, $(\overline{u^2} + \overline{v^2} + 2\overline{uv})/2$, is the component of (normal stress) $\div$ (density) in this direction although in the original axes it is a mixture of normal stresses and a shear stress. As in solids, whether a given arrangement of stresses is "normal" or "shear" or both depends solely on the choice of axes and, again, as in solids, we can always find a set of "principal" axes with respect to which only *normal* stresses remain. The only second-order turbulence quantity that is independent of the axes (i.e. a scalar) is $\overline{u^2} + \overline{v^2} + \overline{w^2}$, written as $\overline{q^2}$ for brevity and to show that it is the mean square of the resultant velocity fluctuation q. The sum of the principal stresses is $-\varrho\overline{q^2}$, and $\frac{1}{2}\varrho\overline{q^2}$ is called the turbulent (kinetic) energy per unit volume. The justification for the name is that we can deduce, from the Navier–Stokes equations, a conservation equation for the value of $\frac{1}{2}\varrho\overline{q^2}$ in the elementary control volume of Fig. 10 and identify the terms with quantities representing sources or sinks of energy or with "transport" of energy in or out of the control volume. The terms appear in groups of three, one for

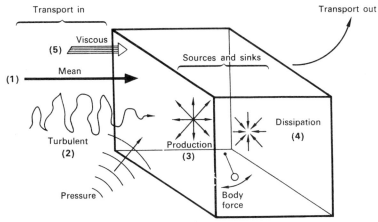

FIG. 10. General form of transport equations—sum of transport terms (spatial gradients of turbulence mean quantities) equals sum of source and sink terms. Numbers refer to discussion in Section 2.2.

each direction: here we give only one term in each group, from which the others follow by interchange of variables, and justify the terms by control volume analysis. The analysis below refers to incompressible flow: density changes produce extra terms. Analogous transport equations can be written for the individual Reynolds stresses (see Appendix 1).

(1) The instantaneous rate of transport of q^2 is q^2 times the instantaneous velocity—e.g. the rate of transport in the x direction is $q^2(U + u)$. The rate of transport of the *mean*, $\overline{q^2}$, in the x direction is therefore $\overline{q^2}U + \overline{q^2 u}$ so that the net rate at which $\tfrac{1}{2}\varrho\overline{q^2}$ is leaving the elementary control volume is the sum of

$$\varrho\frac{\partial}{\partial x}\left(\frac{1}{2}\overline{q^2}\,U + \frac{1}{2}\overline{q^2 u}\right)dx\,dy\,dz$$

and the corresponding terms for the other two directions. The group

$$\frac{\partial}{\partial x}\left(\frac{1}{2}\varrho\overline{q^2}U\right) + \frac{\partial}{\partial y}\left(\frac{1}{2}\varrho\overline{q^2}V\right) + \frac{\partial}{\partial z}\left(\frac{1}{2}\varrho\overline{q^2}W\right)$$

$$\equiv \left(U\frac{\partial}{\partial x} + V\frac{\partial}{\partial y} + W\frac{\partial}{\partial z}\right)\frac{1}{2}\varrho\overline{q^2}$$

is called the *advection* or *mean energy transport* and is the rate of change of $\frac{1}{2}\varrho\overline{q^2}$ along a mean streamline.

(2) The three terms like $(\partial/\partial x)\,(\overline{p'u})\;\mathrm{d}x\,\mathrm{d}y\,\mathrm{d}z$ where p' is the fluctuating part of the pressure, also represent a net loss of turbulent energy from the control volume via the work done in transporting the fluid through a region of changing pressure. It is common to regard this as akin to the terms containing $\overline{q^2 u}$ and to refer to the sum as the *turbulent energy transport* or *energy diffusion*: the instantaneous value of $p' + \frac{1}{2}\varrho q^2$ has some physical significance but it cannot be regarded simply as an instantaneous total pressure because Bernoulli's equation defines total pressure only in the mean.

(3) Terms like $-\varrho\overline{u^2}(\partial U/\partial x)\;\mathrm{d}x\,\mathrm{d}y\,\mathrm{d}z$ represent an extraction of energy from the mean flow by the turbulence: the term quoted is the rate at which work has to be done against the x-component of Reynolds normal stress to stretch an element of fluid $\mathrm{d}x\,\mathrm{d}y\,\mathrm{d}z$ in the x direction at the rate $\partial U/\partial x$. Usually more important in practice are terms like $-\varrho\overline{uv}(\partial U/\partial y)\;\mathrm{d}y\,\mathrm{d}x\,\mathrm{d}z$ which represent work done against the shear stress $-\varrho\overline{uv}$ to shear an element of fluid at the rate $\partial U/\partial y$ [note that there is a second term involving $\overline{uv}$, namely $-\varrho\overline{uv}(\partial V/\partial x)\;\mathrm{d}x\,\mathrm{d}y\,\mathrm{d}z$ for the other constituent of the rate of strain $(\partial U/\partial y) + (\partial V/\partial x)$ so that the sum of the two is $-\varrho\overline{uv}(\partial U/\partial y + \partial V/\partial x)\;\mathrm{d}x\,\mathrm{d}y\,\mathrm{d}z$, containing the product of the Reynolds shear stress and the rate of strain]. From the point of view of the turbulent motion, these are energy *production* terms; from the point of view of the mean motion, they are energy *loss* terms. We must remember that, although the separation of the velocity into mean and fluctuating parts is a perfectly legitimate mathematical device, the instantaneous velocity profile in, say, a pipe may look very different from the mean profile, and there is no difference in kind between the eddy processes that produce, say, $\overline{uv}(\partial U/\partial y)$ and those that produce $\overline{uv(\partial u/\partial y)}$ which is part of the turbulent energy transport in the y direction.

(4) The mean scalar product of the instantaneous *viscous* stress and the instantaneous rate of strain represents a transfer of energy, analogous to the above, between the turbulent motion and the molecular motion— that is, it is a dissipation of turbulent kinetic energy into thermal internal energy by viscosity. From the point of view of the temperature field it is a *source* of energy and is responsible for "aerodynamic heating" in

high-speed flows. The rate of viscous *dissipation* of turbulent energy in the elementary control volume† $\varepsilon \, dx \, dy \, dz$ is the sum of terms like

$$\mu \overline{\left(\frac{\partial u}{\partial y} + \frac{\partial v}{\partial x}\right)^2} dx \, dy \, dz \quad \text{and} \quad \mu \overline{\left(2\frac{\partial u}{\partial x}\right)^2} dx \, dy \, dz$$

(the terms like

$$\mu \left(\frac{\partial U}{\partial y} + \frac{\partial V}{\partial x}\right)^2 dx \, dy \, dz \quad \text{and} \quad \mu \left(2\frac{\partial U}{\partial x}\right)^2 dx \, dy \, dz$$

are direct dissipation of *mean flow* kinetic energy into thermal internal energy: since the instantaneous velocity gradients in the turbulence are much greater than the mean velocity gradients except for the flow extremely close to a solid surface, the viscous dissipation of turbulent energy is usually much greater than the dissipation of mean-flow energy).

(5) Lastly,

$$\mu \left(\frac{\partial^2}{\partial x^2} + \frac{\partial^2}{\partial y^2} + \frac{\partial^2}{\partial z^2}\right) \tfrac{1}{2} \overline{q^2} \, dx \, dy \, dz$$

and terms like

$$\mu \frac{\partial^2}{\partial x \partial y} \overline{uv} \, dx \, dy \, dz$$

represent *viscous transport* of turbulent energy. Like the mean-flow dissipation they are negligible except near a solid surface, and tend to be forgotten altogether in discussions of turbulent energy.

The physical meaning of terms (1) to (5) is summarized in Fig. 10.

The equation obtained by adding up all these terms and equating the sum to zero is the required conservation or "transport" equation for the turbulent energy or the sum of the Reynolds normal stresses. Since it is not an equation we can *solve*, there is no present need to write it out in full: it is written down in simplified form for a two-dimensional thin shear layer in Appendix 1, and is given in full, in tensor notation, in several textbooks [e.g. ref. 5, eqn. (2.4.10)]. Frequently, some of the terms can be neglected: usually, production and dissipation are the

---

† The tensor form is $\dfrac{1}{2}\mu \overline{\left(\dfrac{\partial u_i}{\partial x_j} + \dfrac{\partial u_j}{\partial x_i}\right)^2}$ which looks (but is not) half as large as the sum of the terms quoted here, because each term appears twice on summation.

largest although energy diffusion and advection outweigh them near the edge of a shear layer.

The equation can be regarded as an equation for the rate of change of $\frac{1}{2}\varrho\overline{q^2}$ along a mean streamline in terms of local quantities: the actual value of $\overline{q^2}$ would have to be obtained by integrating the equation along the whole length of the mean streamline, starting at the point where it entered the region of turbulent flow. In non-mathematical language, $\overline{q^2}$ depends on the whole history of the flow, in particular the history of the mean rate of strain which appears in the production terms [(3) above]. The same applies to the separate Reynolds stresses, normal or shear. Only if the transport terms were negligible compared to the local (source and sink) terms could one deduce a direct "eddy viscosity" relation between the Reynolds stresses and the mean rate of strain at the same point. Since the transport terms are often fairly small compared with the source and sink terms, empirical "eddy viscosity" and other "turbulent diffusivity" relations are often fairly good as a first approximation, but it is dangerous to rely on them uncritically.

2.3. Spatial Correlations

The turbulent energy equation contains only mean products of fluctuating quantities at one point in space. To get an idea of the length scales of the fluctuating motion we study the correlation between the same fluctuating quantity measured at two different points in space, for which we can also derive a "conservation" equation from the Navier–Stokes equations (see Section 2.9 for a discussion of more complicated correlations). Suppose that the two points have coordinates (x, y, z) and $(x + r_1, y + r_2, z + r_3)$ or $\mathbf{x}$ and $\mathbf{x} + \mathbf{r}$ in vector shorthand: then the *covariance* of the u-component is defined as $\overline{u(\mathbf{x})\,u(\mathbf{x} + \mathbf{r})}$, a function of $\mathbf{x}$ and of $\mathbf{r}$ in general, and the *correlation coefficient* is

$$\frac{\overline{u(\mathbf{x})\,u(\mathbf{x} + \mathbf{r})}}{\sqrt{\left(\overline{u^2(\mathbf{x})}\cdot\overline{u^2(\mathbf{x} + \mathbf{r})}\right)}}.$$

The dimensionless quantity $\overline{u(\mathbf{x})\,u(\mathbf{x} + \mathbf{r})}/\overline{u^2(\mathbf{x})}$ is preferred to the correlation coefficient in most turbulence work, simply for convenience: it is sometimes written as $R_{11}(r_1, r_2, r_3)$, symbols R_{22} and R_{33} being used for the v and w correlations. Often the covariance itself is called a "correlation". As a great deal of labour is needed to make measurements for all values of $\mathbf{r}$, most of the published work refers to correlations between points separated in one of the coordinate directions: for instance, $R_{22}(r_1, 0, 0)$ is the v correlation with separation r_1 in the x direction (Fig. 11).

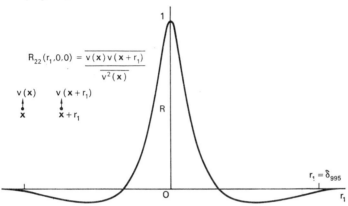

FIG. 11. Typical $R_{22}(r_1, 0, 0)$ correlation in boundary layer at about $y/\delta_{995} = 0.5$.

Physically, the correlation with separation r is a measure of the strength of eddies whose length in the direction of the vector $\mathbf{r}$ is greater than the magnitude of r (since eddies smaller than this will not contribute to the correlation): as a warning that this is an imprecise concept, note that the correlations for the three velocity components will in general be *different* for the same $\mathbf{r}$. The concepts of eddy wavelength are better dealt with by discussing spectra, but the latter are best understood by discussing their mathematical relation to correlations, which we shall arrive at in Section 2.4. When we speak of the length scale of the energy-containing eddies we mean a length of order $\int_0^\infty R \, dr$ (actually called the "integral scale").

2.4. Time Correlations (Autocorrelations or Autocovariances)

The correlation between the same (Greek *autos* = self or same) fluctuating quantity measured at two different *times* (at the same point in space) is not itself very relevant to the behaviour of turbulence and its measurement requires a time delay mechanism (usually a tape-recorder with movable heads or a digital sample-and-delay system): the usual reason for interest is that *if* the turbulent velocity fluctuations are small compared with the mean velocity, the eddies or vortex lines do not change appreciably in shape as they pass a given point and therefore the autocorrelation of the v component (say) with time delay τ, written as $R_{22}(\tau) \equiv \overline{v(t)\,v(t + \tau)}/\overline{v^2}$, will be the same as the space correlation with separation $-U\tau$ in the x direction (which we suppose for simplicity to be the direction of the mean velocity). Correlations with separation in the direction of the mean velocity are difficult to measure except by optical methods because the wake of any probe inserted into the flow is liable to interfere with the flow round the downstream probe, so that measurements of the autocorrelation may be more reliable.

A comparison of the autocorrelation with delay τ and the $(r, 0, 0)$ correlation with separation $-U\tau$ is a measure of the accuracy of the above hypothesis about the slow rate of change of shape of turbulent eddies, a hypothesis due, like so many ideas about turbulence, to G. I. Taylor. Correlations with space *and* time delay are used in more detailed investigations of departures from Taylor's hypothesis in flows with high turbulent intensity. Note that the autocorrelation is always an even function (the same for positive and negative τ) if the turbulence is statistically stationary in time: space correlations may be asymmetrical at large values; an asymmetrical $(r, 0, 0)$ correlation implies that the turbulence is changing appreciably in an x distance equal to the size of the largest eddies, which in turn implies high Reynolds stress and turbulent intensity and, therefore, a departure from Taylor's hypothesis.

2.5. Frequency Spectra

The autocovariance of a quantity that varies as $a \sin \omega_1 t$ is easily seen to be $(a^2/2) \cos \omega\tau$: this autocorrelation of a fluctuation at a single frequency does not decay to zero as $\tau \to \infty$, unlike the typical correlation for turbulence quantities which cover a wide range of frequencies, so there is evidently some connection between the shape of the autocorrelation and the distribution of the fluctuation over different frequencies, the latter being called the spectrum.

A quantity that varies sinusoidally at constant amplitude is said to have a line spectrum, since *any* definition we may choose for the "distribution of the fluctuation over different frequencies", ϕ say, will lead to a concentration of the spectral density ϕ at the frequency ω_1 (Fig. 12). If $u_1 = a \sin \omega_1 t$ and $u_2 = b \sin \omega_2 t$ we get two lines for the spectrum of $u_1 + u_2$: note that $\overline{u_1 u_2} = 0$; fluctuations at different frequencies

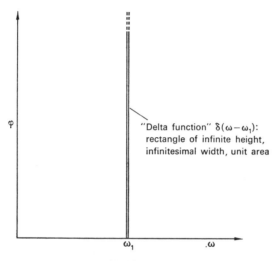

"Delta function" $\delta(\omega - \omega_1)$: rectangle of infinite height, infinitesimal width, unit area

FIG. 12. Line spectrum.

are always uncorrelated. Now we obviously want to define ϕ so that the spectrum of $u_1 + u_2$ is the sum of the spectra of u_1 and u_2 separately: noting that u_1 and u_2 are uncorrelated so that the mean square of $u_1 + u_2$ is the sum of the mean squares of u_1 and u_2 separately, we define the *power spectral density* ϕ by saying that $\phi(\omega) \, d\omega$ is the contribution from frequencies between $\omega - \frac{1}{2} d\omega$ and $\omega + \frac{1}{2} d\omega$ (where $d\omega$ is small but arbitrary) to the mean square of the signal: as pointed out in Section 2.2, root-mean-squares are not useful because they do not add. Mathematically, the spectrum of $a \sin \omega_1 t$ is $(a^2/2) \, \delta(\omega - \omega_1)$, where $\delta(\omega - \omega_1)$ is the delta function, which is zero when $\omega \neq \omega_1$ and goes to infinity at $\omega - \omega_1$ in such a way that $\int \delta(\omega - \omega_1) \, d\omega$ is unity: thus the integral of the spectrum of $a \sin \omega_1 t$ over all ω is the mean square, $a^2/2$. In practice, ϕ can be measured for any fluctuating quantity u with zero mean by putting an electrical signal, proportional to u, through a "band-pass" filter, an arrangement of resonant circuits which, ideally, passes frequencies between $\omega_1 - \tfrac{1}{2} d\omega$ and $\omega_1 + \tfrac{1}{2} d\omega$ only (Fig. 13), and finding the mean square of the output, $\overline{u^2(\omega)}$ say, as if it were a complete signal:ϕ is then $1/d\omega$ times this mean square value.

Turbulent fluctuations have a broad spectrum (Fig. 13): sometimes fairly flat peaks can be distinguished but discrete frequencies occur only in the earlier stages of transition from laminar to turbulent flow. The

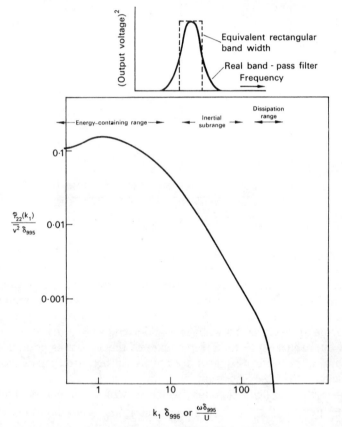

FIG. 13. Typical $\phi_{22}(k_1)$ wave number (or frequency) spectrum in boundary layer at about $y/\delta_{995} = 0.5$.

phase of the contribution to the spectrum at a given frequency itself fluctuates randomly. A good deal of work has been done on the mathematics of random spectra but, as usual, the most helpful definitions are embodied in the measurement technique: in case of doubt, ask "what does a band-pass filter measure?"

There is a simple connection between the spectral density $\phi(\omega)$, as defined above, and the covariance $\overline{u(t)\,u(t + \tau)}$. For the present purpose we can illustrate it by asking how we can deduce the autocovariance of a signal $a \sin \omega_1 t$ [which we know to be $(a^2/2) \cos \omega_1 \tau$] from its spectrum [which we know to be $(a^2/2)\,\delta(\omega - \omega_1)$]. A little thought will show that *all* frequencies of the signal will contribute to the autocorrelation for given τ, and vice versa, so that the autocorrelation must be a weighted integral of ϕ over the whole range of ω. A little more thought shows that the obvious weighting factor is $\cos \omega\tau$, because $\int F(\omega)\,\delta(\omega - \omega_1)\,d\omega$ over all ω is $F(\omega_1)$ for *any* function F: in particular, for $\tau = 0$ we get $\int \phi\,d\omega$, which is $\overline{u^2}$, the value of $\overline{u(t)\,u(t + \tau)}$ at $\tau = 0$. This is not, of course, a proof that

$$\overline{u(t)\,u(t + \tau)} = \int\limits_0^\infty \phi(\omega) \cos \omega\tau \, d\omega$$

for *any* spectrum function ϕ, where the integral extends over all frequencies that contribute to ϕ: nevertheless, this relation can be proved and is an example of a *Fourier transform*. The most general Fourier transform has a weighting factor

$$\exp\,(\pm i\omega\tau) \equiv \cos \omega\tau \pm i \sin \omega\tau$$

where $i = \sqrt{-1}$: all the physical quantities we have to deal with are real, and complex numbers are used, as always in mathematical physics, simply as devices to identify different parts of the transform and to exploit the useful properties of $\exp\,(i\omega\tau)$. The general transform deals with functions which, unlike $\phi(\omega)$, are not necessarily even functions (i.e. not necessarily taking equal values for equal positive and negative values of the argument) and which in general cover the whole range of the argument from $-\infty$ to ∞: therefore the limits of integration in the above integral are really $-\infty$ and ∞, which raises the question of negative frequencies. Strictly, we should regard ϕ as an even function of ω, extending over both positive and negative frequencies so that, for instance, the spectrum of $a \sin \omega_1 t$ should really be $a^2/4[\delta(\omega - \omega_1) + \delta(\omega + \omega_1)]$: however, if the time correlation is an even function of τ we can simplify matters by supposing that the spectrum contains positive frequencies only, so that

$$\int_0^\infty \phi(\omega) \, d\omega = \overline{u^2} \quad \text{and} \quad \int_0^\infty \phi(\omega) \cos \omega\tau \, d\omega = \overline{u(t) \, u(t + \tau)}$$

for ϕ as defined near the beginning of this section. It can be shown that the inverse transform

$$\phi(\omega) = \frac{1}{\pi} \int_{-\infty}^\infty \overline{u(t) \, u(t + \tau)} \cos \omega\tau \, d\tau = \frac{2}{\pi} \int_0^\infty \overline{u(t) \, u(t + \tau)} \cos \omega\tau \, d\tau$$

also holds: again, the contribution to $R(\tau)$ at negative τ can be ignored. The factor $2/\pi$, or $1/2\pi$ if we are rigorous and recognize negative arguments in correlation *and* spectrum, has to appear somewhere in the transform or the inverse transform for reasons of consistency (see ref. 2) but its position can be chosen to suit our convenience: since we want $\int \phi(\omega) \, d\omega$ to be the mean square of the quantity considered, we put the factor in the inverse transform.

The practical use of Fourier transforms is that we can base our experiments—or our theory—on *either* the spectrum or the correlation, whichever is more convenient: they contain exactly the same information. We can measure either the frequency spectrum or the autocorrelation fairly easily but we cannot directly measure the so-called wave number spectrum, which is the Fourier transform, with respect to **r**, of the *space* correlation: since the wave number spectrum is usually more convenient than the space correlation for theoretical purposes, Fourier transform relations are invaluable in this case.

2.6. Wave Number Spectra

2.6.1. *Derivation*

It is natural to ask the significance of the Fourier transform of the *space* correlation, say for simplicity $R_{22}(r_1, 0, 0)$, the v-component correlation with separation r_1 in the x direction, which according to Taylor's hypothesis is nearly the same as the v-component autocorrelation with time delay $-r_1/U$ (or equivalently $+r_1/U$). This transform is called the *wave number spectrum*: in the case mentioned it is the one-dimensional (x-component) wave number spectrum of the v-component of velocity.

If $v = a \sin k_1' x$, neglecting time dependence for simplicity, the space correlation is $\tfrac{1}{2} a^2 \cos k_1' r_1$, and the wave number spectrum is $\tfrac{1}{2} a^2 \delta(k_1 - k_1')$ for $k_1 > 0$, writing the x-component wave number as k_1 to show that it goes with r_1 (y and z component wave numbers would be k_2 and k_3 respectively; this is standard notation): thus a sine wave of wavelength λ has a line spectrum at wave number $2\pi/\lambda$, and the wave number k bears the same relation to the wavelength $\lambda \equiv 2\pi/k$ as the radian frequency ω bears to the period $2\pi/\omega$.

The Fourier transform relation between the above space correlation and the k_1 wave number spectrum must be written in the general form,

$$\phi_{22}(k_1) = \frac{1}{2\pi} \int_{-\infty}^{\infty} \overline{v(\mathbf{x}) \, v(\mathbf{x} + r_1)} \exp(-ik_1 r_1) \, dr_1,$$

$$\overline{v(\mathbf{x}) \, v(\mathbf{x} + r_1)} = \int_{-\infty}^{\infty} \phi_{22}(k_1) \exp(ik_1 r_1) \, dk_1,$$

so that if the correlation has an odd part as well as an even part ϕ will have an imaginary part as well as a real part. As mentioned above, the complex notation is used only to distinguish the two parts of the spectrum: most of the elementary use made of wave number spectra ignores the imaginary part, which is usually small because the correlation is usually nearly even. Sometimes $\phi(k_1)$, like $\phi(\omega)$, is defined for positive k only.

To add the final stage of complication we may define a *three-dimensional wave number spectrum*, by an obvious extension of the last equation but one, as

$$\Phi(k_1, k_2, k_3) = \frac{1}{(2\pi)^3} \iiint \overline{v(\mathbf{x}) \, v(\mathbf{x} + \mathbf{r})} \exp(-ik_1 r_1) \exp(-ik_2 r_2) \times$$

$$\times \exp(-ir_3 r_3) \, dr_1 \, dr_2 \, dr_3$$

with the inverse

$$\overline{v(\mathbf{x})\, v(\mathbf{x} + \mathbf{r})} = \iiint \Phi(k_1, k_2, k_3) \exp(ik_1 r_1) \exp(ik_2 r_2) \times$$

$$\times \exp(ik_3 r_3)\, dk_1\, dk_2\, dk_3$$

where the limits on the integrals are $-\infty$ and ∞ in each case.

The three exponentials combine into $\exp[i(k_1 r_1 + k_2 r_2 + k_3 r_3)]$ sometimes written as $\exp(i\mathbf{k} \cdot \mathbf{r})$. This introduces the *wave number vector* $\mathbf{k}$, which has components k_1, k_2, k_3 and is the analogue of the separation vector $\mathbf{r}$ which has components r_1, r_2, r_3: it is convenient to talk of the distribution of Φ in "wave number space", which has coordinates k_1, k_2, k_3. Putting $r_2 = 0$, $r_3 = 0$ in the equation for $\overline{v(\mathbf{x})\, v(\mathbf{x} + \mathbf{r})}$ gives

$$\overline{v(\mathbf{x})\, v(\mathbf{x} + r_1)} = \int \left(\iint \Phi(k_1, k_2, k_3)\, dk_2\, dk_3 \right) \exp(ik_1 r_1)\, dk_1$$

and comparison with the equation for $\overline{v(\mathbf{x})\, v(\mathbf{x} + r_1)}$ (p. 35) shows that

$$\phi(k_1) = \iint \Phi(k_1, k_2, k_3)\, dk_2\, dk_3 \quad \text{not} \quad \Phi(k_1, 0, 0).$$

It is a useful mnemonic to represent the distribution of $\Phi(\mathbf{k})$ in wave number space by a solid body, whose local density is equal to the local value of Φ (Fig. 14): then the weight of a slice of thickness dk_1, cut perpendicular to the k_1-axis, is $\phi(k_1)\, dk_1$. Another degenerate form, related to the energy spectrum $(1/2)(\Phi_{11} + \Phi_{22} + \Phi_{33})$, is $E(k)$, where $2E(k)\, dk$ is the weight of a spherical shell, of radius k and thickness dk, cut from the imaginary solid: $E(k)$ is called the wave-number-magnitude spectrum and is related to $\Phi(\mathbf{k})$ by

$$E(k) = (1/2) \int (\Phi_{11}(\mathbf{k}) + \Phi_{22}(\mathbf{k}) + \Phi_{33}(\mathbf{k}))\, dA$$

where the integral is taken over the surface of the sphere of radius k.

The only practicable way of obtaining true wave number spectra experimentally is as the Fourier transforms of the corresponding correlations: one can devise multiple arrays of measuring instruments which respond only to a given band of wave numbers but these require large numbers of instruments strung out in the direction of the required

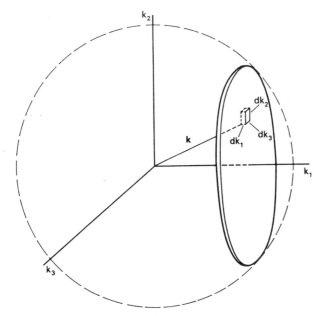

Energy content of cube $= \Phi\,(\mathbf{k})\,dk_1\,dk_2\,dk_3$

Energy content of slice $= dk_1 \iint \Phi\,(\mathbf{k})\,dk_2\,dk_3 = \varphi\,(k_1)\,dk_1$

FIG. 14. One-dimensional and three-dimensional spectra.

wave number component and amount to a semimechanical method of Fourier-transforming the correlation. Fortunately $\phi(\omega)$ is a good enough approximation to $\phi(k_1)$ for most purposes.

2.6.2. *Physics*

Most of the basic theoretical work on turbulence (e.g. ref. 2) is concerned with the behaviour of wave number spectra and with the transfer of turbulent energy from low wave numbers (large wavelengths) to the high wave numbers (small wavelengths) where it is dissipated (see Section 1.6 and Fig. 13). If the ratio of the wave numbers in the dissipating range to those in the energy-containing range is high enough (a factor of 10 seems minimal) the eddies in the dissipating range will be isotropic and independent of the details of the energy-containing range, which may be significantly anisotropic, and the structure in the dissipating range will depend only on the energy transfer from the energy-containing range (equal to the rate of energy dissipation ε), and on the wave number, viscosity and density. Typical length and velocity scales for the dissipating eddies, called the Kolmogorov scales, are obtainable from dimensional analysis as $\eta \equiv (v^3/\varepsilon)^{\frac{1}{4}}$ and $v = (v\varepsilon)^{\frac{1}{4}}$ (the fact that the Reynolds number $v\eta/v$ is unity is just a consequence of the dimensional analysis: actually the dissipating range begins at about $k = 0\cdot1/\eta$ and ends at about $k = 1/\eta$ so that wavelengths in the dissipating range are 6η to 60η). Now according to elementary kinetic theory the molecular mean free path λ in a gas can be written as $\sqrt{(\gamma\pi/2)}\ v/a$ where a is the speed of sound: it follows that $\lambda/\eta = \sqrt{(\gamma\pi/2)}$ $\times\ v/a$ so that the ratio of mean free path to Kolmogorov length scale is about 1·5 times the Mach number based on v. Since this is always small compared to a Mach number based on, say, $\sqrt{\overline{u^2}}$, which is in turn always small compared to a Mach number based on mean velocity in a shear flow, we can always neglect departures from continuum flow if the Reynolds number is high enough for turbulence to exist. Dimensional analysis shows that the one-dimensional spectral density $\phi(k_1)$ or the wave-number magnitude spectral density $E(k)$, both of which have the dimensions of (energy)/(wave number), have the form

$$v^2\eta f(k\eta) = v^{\frac{5}{4}}\varepsilon^{\frac{1}{4}}f\left(k\left(\frac{v^3}{\varepsilon}\right)^{\frac{1}{4}}\right)$$

within the isotropic range only. If the dissipating wave numbers are really high compared with the energy-containing wave numbers (a factor of 100 seems minimal) then the conditions for isotropy, and for independence of the details of the energy-containing eddies, may also be satisfied for a range of wave numbers below the dissipating range, where the eddy structure will be independent of viscosity: this range is called the inertial subrange, and the spectrum depends only upon k and the energy transfer through wave number k (equal to ε). Therefore, by dimensional analysis, the spectra in this subrange are

$$E(k) = \alpha\varepsilon^{\frac{2}{3}}k^{-\frac{5}{3}},$$

$$\phi(k_1) = K'\varepsilon^{\frac{2}{3}}k_1^{-\frac{5}{3}}.$$

It can be shown that $K' = 18\alpha/55$ for the u component spectrum as defined for $k > 0$ only, and 4/3 times this for the v and w spectra (although the turbulence is isotropic the spectra of a given component depend on the angle between the component and the wave number chosen). The universal constants, α and K', must be found experimentally unless they can be predicted by a theory: currently accepted values are about 1·5 and 0·5, respectively. If the dissipation range is isotropic, the dissipation rate ε simplifies to $15\nu\overline{(\partial u/\partial x)^2}$ (see Appendix 1) which can be deduced from measurements of $\overline{(\partial u/\partial t)^2}$ by invoking Taylor's hypothesis. The length $[\overline{u^2}/\overline{(\partial u/\partial x)^2}]^{\frac{1}{2}} \equiv \lambda$, and the analogous lengths defined in terms of the other components, are sometimes called the *microscales*: they are far larger than the length scale η of the dissipating eddies.

For instance, in a shear layer of thickness δ, ε is of order $(\overline{u^2})^{3/2}/\delta$ so λ/η is of order $[(\overline{u^2})^{\frac{1}{2}}\delta/\nu]^{\frac{1}{2}}$ if the dissipation is isotropic. The equalities

$$\overline{(\partial u/\partial x)^2} = -\frac{1}{2}\overline{u^2}\left[\frac{d^2}{dr_1^2}R_{11}(r_1,0,0)\right]_{r_1=0} = \int k_1^2\phi_{11}(k_1)\,dk_1,$$

or the corresponding relations for time and frequency, are sometimes useful. Another useful deduction from the Fourier transform relation between ϕ and R is

$$\phi_{11}(k_1 = 0) = \frac{\overline{u^2}}{2\pi}\int_{-\infty}^{\infty}R_{11}(r_1,0,0)\,dr_1,$$

where $\int_0^\infty R\,dr$ is the "integral scale" L which is frequently used as a typical length scale of the energy-containing eddies. It must be emphasized that these various length scales are *typical* scales, for use in order-of-magnitude arguments. For instance, the main contributions to the energy may be spread over a range of length scales of much more than $10:1$, with the position of L in this range depending strongly on the type of flow considered, and obviously even the dissipating eddies contain *some* energy: by "energy-containing" and "dissipating" ranges of eddy size we mean those that contain *most* of the energy or effect *most* of the dissipation.

Order-of-magnitude arguments are often very useful in turbulence and are just as reliable as exact analyses for the purposes for which they are intended: one is much more likely to suffer trouble from inaccurate physical premises than from inexact order-of-magnitude analyses. An example is the simple physical argument, used above, that the smaller eddies will depend only on the mean energy transfer from the energy-containing eddies and not on the statistical properties of the latter. This cannot be quite correct: when discussing the smaller eddies we can define the "mean" energy transfer by an average over, say, 100 wavelengths or periods of the smaller eddies, and this average will itself vary at the much larger wavelength or period of the energy-containing eddies, so that the properties of the small-scale eddies have a "spotty" distribution in space and time. According to plausible theoretical estimates, this does not greatly affect the simple dimensional results given above—for instance, the $-5/3$ ($-1\cdot67$) power law is replaced by a $-1\cdot71$ power law—but the phenomenon has to be borne in mind when considering the smaller-scale motion in more detail.

A final warning about the use of spectra concerns the relation between one-dimensional and three-dimensional spectra. Since $\phi(k_1)$ is the result of an integration of $\Phi(\mathbf{k})$ over all values of k_2 and k_3, care is needed in discussing its physical significance. In particular, the value of ϕ near $k_1 = 0$ arises mainly from eddies or elementary vortices with comparatively high k_2, k_3 wave numbers, as they pass through the k_1 direction: Φ is zero at $\mathbf{k} = 0$. At high wave numbers, where Φ and ϕ both fall rapidly with increasing wave number, most of the contributions to the integral for ϕ come from near $\mathbf{k} = k_1$ (see Fig. 14), and the correspondence between Φ and ϕ is closer.

2.7. Space–Time Correlations

The study of departures from Taylor's hypothesis of quasi-rigid convection involves correlations with separation both in space and in time pioneered by Favre at Marseille. An obvious way of studying the development of eddies as they are swept downstream is to compare space–time correlations, having time separation τ and space separation $r_1 + U\tau$ in the direction of the mean velocity, with the original space correlation $R(r_1, 0, 0)$: if the eddies really were rigidly convected for a distance $U\tau$ in time τ the two correlations would be identical but, in fact, (Fig. 15), the delayed correlation is weaker because the eddies are changing all the time, even if statistical mean quantities are independent of x.

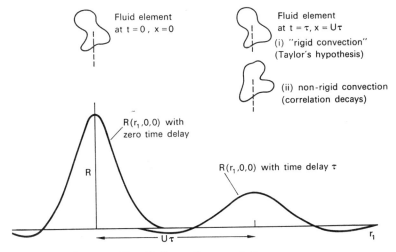

Fluid element
at $t = 0$, $x = 0$

Fluid element
at $t = \tau$, $x = U\tau$
(i) "rigid convection"
(Taylor's hypothesis)

(ii) non-rigid convection
(correlation decays)

$R(r_1, 0, 0)$ with
zero time delay

$R(r_1, 0, 0)$ with time delay τ

R

$U\tau$

r_1

FIG. 15. Space–time correlation.

The speed at which turbulent energy is convected in the x direction is

$$\left(\frac{1}{2} \overline{q^2(U + u)} + \frac{\overline{p'u}}{\varrho} \right) \bigg/ \frac{1}{2} \overline{q^2}$$

according to the analysis of Section 2.2. This reduces to

$$U + \left(\frac{\overline{p'u}}{\varrho} + \frac{1}{2} \overline{q^2 u} \right) \bigg/ \frac{1}{2} \overline{q^2}$$

and is one of the many possible definitions of the "convection velocity" of the turbulence. Although we have neglected the viscous diffusion of turbulent energy in the x direction, and although apart from this effect the vortex lines move with the fluid, the preferential stretching of vortex lines in different directions can produce a net transport of vorticity or turbulent energy relative to the fluid (another example of net energy transfer without mass transfer is the propagation of sound waves). Clearly, the velocity at which energy is transported relative to the fluid due to vortex stretching will be of the same order as the velocity fluctuations themselves and, therefore, a fairly small fraction of the local mean velocity in typical turbulent flows. However, the term $\overline{p'u}$ represents a convection of energy associated with pressure fluctuations: the pressure fluctuations are generated by the turbulence, it is true, but not necessarily the turbulence near the point considered, because mean or fluctuating pressures can influence the whole of space according to an inverse square law; the convection velocity of the pressure fluctuations is—roughly—the fluid velocity in the region where they were generated, which may be different from the velocity at the point where $\overline{p'u}$ is measured. As well as the above "hydrodynamic" or "near-field" pressure fluctuations, which arise whether the fluid is compressible or not and which move, in effect, with the fluid that generates them, the turbulence may generate sound waves which propagate to infinity at the speed of sound in directions that depend on the orientation of the noise-producing turbulence: the acoustic energy propagation rate is also contained in the $\overline{p'u}$ term. The difference between true sound, hydrodynamic (as opposed to acoustic) pressure fluctuations and vorticity fluctuations is most simply remembered by thinking of a fast-moving vehicle passing an observer, who feels velocity fluctuations caused first by the noise, then by the static pressure field around the vehicle and, finally, by the turbulent wake. The same three effects contribute to the space–time correlation in a turbulent field: the contribution of sound waves is negligible at low Mach numbers except perhaps for noise produced by the test rig itself, but the hydrodynamic pressure fluctuations produce appreciable "irrotational" velocity fluctuations.

The usual reason for studying the convection velocity, apart from its fundamental interest, is in connection with the excitation of running waves in a liquid or solid surface below a turbulent stream: this is most conveniently studied in terms of the wave number/frequency spectrum which is the complete Fourier transform of the space–time correlation (usually for one-dimensional wave number, giving contours of equal spectral density in the k_1, ω-plane; this can be further modified into contours in the $k_1, \omega/k_1$-plane, where ω/k_1 is a typical phase velocity, and any rational definition of convection velocity can be quickly extracted).

The details of the development of the eddies viewed in a frame of reference moving with a velocity that minimizes their rate of change with time (one of the possible definitions of convection velocity) are of interest in examining noise production by turbulence: the equations governing noise production can be derived from the Navier–Stokes equations, but even the process of extracting an expression for the mean-square sound pressure fluctuation at a point far from the turbulence is not straightforward and the turbulence correlations that appear on the right-hand side of the final expression are very complicated and rather ill-conditioned (turbulence is, fortunately, a very inefficient generator of sound: of course, this means that attempts to reduce the efficiency of sound generation even more are unlikely to succeed).

2.8. Cross-correlations and Cross-spectra

Clearly, we can, if we wish, measure the correlation between *different* components of velocity: for instance, $\overline{u(t)\, v(t + \tau)}$ is the time-delayed correlation between u and v. Its Fourier transform is the frequency spectrum of $\overline{uv}$, which can be measured directly by putting the u and v signals through separate but identical band pass filters† and taking the mean product of the two outputs, $u(\omega)$ and $v(\omega)$, say. People who think in terms of r.m.s. signals occasionally feel that since the spectrum $\phi(\omega)$ of $\sqrt{\overline{u^2}}$ is $\overline{[u(\omega)]^2}$ the spectrum of $\overline{uv}$ should be $\overline{u^2(\omega)\, v^2(\omega)}$: however, one must remember that the u-component spectrum is the frequency distribution of contributions to $\overline{u^2}$, not $\sqrt{\overline{u^2}}$. Since $\overline{u(\mathbf{x})\, v(\mathbf{x} + \mathbf{r})}$ is not in general an even function of $\mathbf{r}$, it is not equal to $\overline{u(\mathbf{x} + \mathbf{r})\, v(\mathbf{x})}$ and the two must be measured separately; in special cases symmetries can be distinguished. Wave number spectra can be derived as the Fourier transforms of these and other cross-correlations.

2.9. Higher-order Correlations and Spectra

Correlations between three components of velocity at three different points can also be defined: those that appear in turbulence theory are usually degenerate [e.g. those like $\overline{u(\mathbf{x})\, v(\mathbf{x})\, v(\mathbf{x} + \mathbf{r})}$ transform into spectra giving the energy transfer from low wave numbers to high wave numbers]. Correlations of the fourth and higher orders are bound to be partly degenerate because there are only three velocity components to choose from. These correlations appear in "conservation" equations

† *Both* signals must be filtered because a band-pass filter produces a phase shift.

derived from the Navier–Stokes equations by multiplying the latter by products of the velocities at different points before time averaging. For instance, the turbulent energy equation can be obtained by multiplying the x-component Navier–Stokes equation by the u-component velocity at the *same* point, adding the corresponding y and z equations to it and then taking the time average. If the Navier–Stokes equations at one point x are multiplied by velocity components at *another* point $x + r$ we obtain an equation for the rate of change of a correlation with x or r: the simplest equation of this sort is the Karman–Howarth equation for decaying isotropic turbulence[2,5] where the "rate of change" is strictly with respect to time. These equations are for second-order correlations, but contain third-order correlations on the right-hand side (just as the turbulent energy equation contains $\overline{q^2 v}$): if we derive the equations for the third-order correlations by multiplying the Navier–Stokes equations by second-order velocity products, we find fourth-order correlations on the right-hand side, and so on. The infinite set of correlation equations so obtained contain together all the statistical information obtainable from the Navier–Stokes equations: there have been many attempts to truncate the series by making some plausible assumption about the Nth order correlations; so far N has never exceeded 5! For alternative approaches, see Appendix 2.

2.10. Probability Distributions and Intermittency

The various mean products and space correlations are integral moments of the *joint probability distribution* for the velocity components at different points in space. The probability distribution for the single variable u_1 at a single point, $P(u_1)$, is defined (Fig. 9) by saying that $P(u_1)\,du_1$ is the fraction of the total duration of a long sample for which the variable lies between $u_1 - \frac{1}{2}du_1$ and $u_1 + \frac{1}{2}du_1$, i.e. the probability that the variable lies between $u_1 - \frac{1}{2}du$ and $u_1 + \frac{1}{2}du_1$. The joint probability for variables $u_1, u_2, ..., u_n$ is $P(u_1, u_2, ..., u_n)$ where

$$P(u_1, u_2, ..., u_n)\,du_1\,du_2, ..., du_n$$

is the probability that the first variable lies between $u_1 - \frac{1}{2}du_1$ and $u_1 + \frac{1}{2}du_1$ at the same time that the second variable lies between $u_2 - \frac{1}{2}du_2$ and $u_2 + \frac{1}{2}du_2$, the third variable between $u_3 - \frac{1}{2}du_3$ and

$u_3 + \frac{1}{2} du_3$, etc. The n variables can be at different (fixed) points, and P is different for each arrangement of points, so it must be written as

$$P(u_1, u_2, ..., u_n; x_1, x_2, ..., x_n).$$

For a fluctuation velocity u_1 the single probability distribution $P(u_1)$ satisfies

$$\int_{-\infty}^{\infty} P \, du_1 = 1$$

(the velocity must be somewhere between $-\infty$ and ∞),

$$\int_{-\infty}^{\infty} u_1 P \, du_1 = \bar{u}_1 (= 0 \text{ usually})$$

(the mean of a quantity is the integral of the probability weighted by the function itself),

$$\int_{-\infty}^{\infty} u_1^n P \, du_1 = \overline{u_1^n}$$

[$P(u_1')$ is also the probability that u_1^n takes a certain value $u_1'^n$].

The joint probability distribution $P(u_1, u_2)$ for two variables u_1 and u_2 satisfies

$$\int_{-\infty}^{\infty} P(u_1, u_2) \, du_1 = P(u_2),$$

$$\int_{-\infty}^{\infty} \int_{-\infty}^{\infty} P(u_1, u_2) \, du_1 \, du_2 = 1,$$

$$\int_{-\infty}^{\infty} \int_{-\infty}^{\infty} P(u_1, u_2) u_1^m u_2^n \, du_1 \, du_2 = \overline{u_1^m u_2^n}$$

and so on for higher orders of joint probability.

In the simpler cases, it is only the integral moments like $\overline{u_1^n}$ that interest us directly, and these are better measured by multiplication and averaging of the signals rather than by forming the weighted integrals of a measured probability distribution—although this depends on the relative accuracy of the apparatus used for multiplying and that used for probability analysis. For more refined attempts to establish the interaction between different parts of a turbulent flow, probability concepts and the use of P as a weighting function can be used to great effect: the best-known example is the concept of *intermittency*. The free edge of a turbulent flow is quite well defined, but highly irregular and unsteady (see Frontispiece) and we can learn something about the turbulent eddies—generally the larger ones—that produce this irregularity, by studying the probability that the flow at a given point is turbulent: this probability is called the intermittency (1 for fully turbulent flow, 0 for a point that is always outside the turbulence). We can define weighted mean values within the turbulent flow only, instead of mean values taken over all time. In practice we have to define the intermittency as the fraction of time for which the flow at the point considered has a fluctuating vorticity (or some approximation thereto) numerically greater than a small threshold value. If we take a threshold value that is *not* small then we can select the more intense patches of turbulence (near the free edge or elsewhere) and establish the probability distribution of these or obtain weighted mean values within them. These concepts of selecting specially defined regions of turbulence for statistical analysis are in an early stage of development: they arise from attempts to probe more delicately into the mechanism of turbulence than one can do with the blunter instruments like correlations and spectral analysis, but although any weighted mean values we may define form a proper subset of all the integral moments of

$$P(u_1, u_2, ..., u_n; x_1, x_2, ..., x_n)$$

they may not be easily identifiable with terms in an equation derivable from the Navier–Stokes equations, so that we must take care to understand the physical significance of the mean quantities we define.

CHAPTER 3

Examples of Turbulent Flows[5]

... It next will be right
To describe each particular batch:
Distinguishing those that have feathers, and bite,
From those that have whiskers, and scratch.

THE examples given in this chapter, ranked in order of increasing complication, are mainly idealized flows: the real turbulent flows found in nature or industry may be more complicated, but they can often be treated as combinations or distortions of ideal flows (if they cannot be so treated neither mathematics nor experiment is likely to achieve results of general use in the immediate future). Therefore the flows described below are to be regarded as building bricks for the synthesis of more complicated flows, as well as simple educational examples. Experimental results in idealized shear flows are reviewed by Townsend,[5] and I have given references to more recent research reports in the "Further Reading" list so that readers who are interested in quantitative results for particular flows can study them in context. The present book is not the place for a critical analysis of experimental results and I am unwilling to present the results without such analysis.

3.1. Turbulence behind a Grid of Bars[2] (Fig. 16)

The wakes of the individual bars become turbulent close behind the grid and then interact in some very complicated way so that, at a large number of mesh lengths from the plane of the grid, the turbulence is more or less homogeneous. The turbulent energy decays slowly as the distance from the grid increases because there is no source of turbulent

47

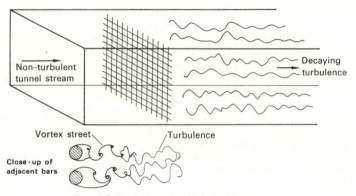

FIG. 16. Turbulence behind a grid of bars.

energy other than the shear flow close to the grid. As the turbulence decays, the energy transfer from the large eddies to the small eddies decreases so that the intensity of the small eddies decreases faster than that of the large eddies. Therefore, the dissipation rate decreases and the dissipating eddies become larger [the typical wavelength of the dissipating eddies being proportional to $(\nu^3/\varepsilon)^{\frac{1}{4}}$]. Finally, the intensity and the dissipation rate sink to such low levels that the dissipation in the largest eddies (the size of the original eddies generated close to the grid) becomes relatively important: this is turbulence of very low Reynolds number (based on $\sqrt{\overline{u^2}}$ and the microscale λ, say), whose behaviour can be predicted by ignoring the inertia terms in the equations of motion and which does not, therefore, throw very much light on the behaviour of turbulence at more normal Reynolds numbers.

    The theoretical idealization of grid turbulence is homogeneous isotropic turbulence (see Glossary) and most of the basic theoretical work on predicting spectrum and correlation shapes has been done in isotropic turbulence[2], but it is fair to say that this work is not currently of much help in calculating the behaviour of shear flows, which have a very much more complicated energy balance. In isotropic turbulence, not only is there no extraction of energy by the turbulence from the mean flow (even if the mean velocity gradients are non-zero, although in such a case the turbulence would immediately depart from isotropy), but there is no lateral transport of turbulent energy by the turbulence itself (energy diffusion). Although energy diffusion is not simply related to the mean gradient of the turbulent energy, it will clearly be zero when the energy and length scales are everywhere the same: the longitudinal diffusion of turbulent energy in grid turbulence, with its very small gradient of turbulent energy in the x direction, can almost always be neglected. A more important shortcoming of grid turbulence is that it is not quite isotropic—the ratios of the turbulent intensities being $0.75\overline{u^2} = \overline{v^2} = \overline{w^2}$ approximately. It can be assumed that eventually the vortex-stretching processes within the turbulence would equalize the three components of turbulent intensity, but this process of return to isotropy seems to be very slow indeed and the anisotropy of grid turbulence is not sufficiently large for the phenomenon to be investigated experimentally. It is curious that vortex stretching produces isotropy with increasing wave number much more effectively than with increasing time.

3.2. "Infinite" Shear Flow[6]

If a constant velocity gradient, $\partial U/\partial y$ say, is superimposed on initially isotropic turbulence, a shear stress $-\varrho\overline{uv}$ will be developed, initially increasing proportional to x/U, and as a result the turbulence will be able to extract energy from the mean flow. The flow will remain homogeneous, though not isotropic, and turbulent energy diffusion will still be negligible. Therefore, this is the simplest type of shear flow, although it is rather difficult to set up experimentally, and some progress has been made in extending the ideas of isotropic turbulence theory to this case. One of the unresolved questions is that of the development of eddy length scale: since there is no mean-flow length scale it seems likely that the energy-containing eddies must either increase to infinite length scale or decrease to infinitesimal length scale. Since for a given turbulent intensity the latter possibility would imply infinite turbulence dissipation rate but only a finite rate of energy production, it can be discounted, and it appears that the length scale must increase monotonically to infinity. The small amount of experimental data available supports this conclusion but, in practice, it is difficult to obtain at the same time a sufficiently high Reynolds number and a sufficiently small ratio of eddy size to test rig size. In any real shear flow an upper limit is set to the size of the turbulent eddies by the dimensions of the mean flow itself, and indeed this upper limit is nearly always reached, as in a pipe flow where the largest eddies have wavelengths of the order of the pipe diameter.

3.3. Couette Flow[7] (Fig. 17)

This is the flow between one fixed and one moving wall. The experimental difficulties are obvious and, like the two preceding flows, this one is used mainly as a theoretical battleground and as a demonstration of the effects of constraint on the development of the turbulent flow. Clearly the eddy size in the y direction cannot exceed the height h of the channel, and the typical wavelengths in the other two directions cannot be an order of magnitude greater than this if the eddy is both

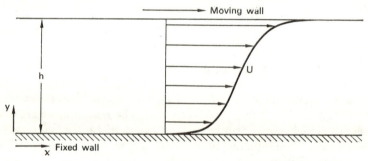

FIG. 17. Velocity profile in turbulent Couette flow.

to satisfy the continuity equation and to produce an appreciable Reynolds stress (writing $\lambda_x, \lambda_y, \lambda_z$ for the wavelengths in the three directions so that $\partial u/\partial x$ is roughly proportional to u/λ_x and so on, continuity requires u/λ_x, v/λ_y and w/λ_z to be of the same order: if $\lambda_x \gg \lambda_y$, $v \ll u$ and $\overline{uv} \ll \overline{u^2}$. This sort of argument is frequently useful in testing hypotheses about eddy shapes and is re-used in the next paragraph).

The total (viscous plus turbulent) shear stress in the x, y-plane, τ, is constant throughout this idealized flow, the shear stress applied by the fluid to the top wall being equal to the stress applied to the bottom wall by the fluid. The velocity gradient in turbulent Couette flow is not constant (although in laminar Couette flow, it is). The mean velocity and all the fluctuating velocities (measured with respect to the wall) are zero at either wall and V is zero everywhere.

We will now derive some very important results for the flow near the bottom (or top) wall. It will be seen later that the assumptions and results apply equally well to the near-wall region of other types of turbulent flow.

Since $\partial u/\partial x$ and $\partial w/\partial z$, and therefore (using the continuity equation) $\partial v/\partial y$, are zero at the wall, the v component of fluctuating velocity increases as the square of the distance from the wall (u and w both increase linearly from the wall, as does the mean velocity U). These facts imply that (1) the ratio of the u and w components to the mean velocity U is not zero at the wall: in fact the ratio of the r.m.s. u component to the mean velocity reaches values of the order of 0·3 at the wall; (2) the Reynolds shear stress $-\varrho\overline{uv}$ must vary at least as rapidly as y^3 very close to the wall. The latter fact has led to the description of the region very close to the wall as the "laminar sublayer": certainly, the viscous shear stress $\mu\,\partial U/\partial y$ is very much larger than the Reynolds shear stress but the presence of large values of $\sqrt{\overline{u^2}}/U$ has led to the name "linear sublayer" (Fig. 18) being preferred. From the same arguments as those applied to the flow as a whole, we can see that a typical length dimension of the turbulent eddies normal to the wall will be of the order of the distance from the wall considered. Therefore, close enough to the wall, the turbulent eddies will be so small that the larger eddies in the outer part of the flow will appear almost steady from their point of view, which means that the statistical properties of *any* turbulent flow near a wall ought to depend only on local variables (note the similarity of this line of reasoning to that used to derive the Kolmogorov scales for turbulence at high wave numbers). The only quantities that appreciably influence the flow close to the solid surface are expected to be the shear stress at the wall, τ_w, the fluid properties, say ϱ and ν, and the distance from the wall itself. Of course, if the wall were rough or if there were a flow of fluid through the wall, then extra quantities would appear. Dimensional analysis shows us that any quantity having the dimensions of velocity (for instance, a mean velocity or a root-mean-square turbulence fluctuation) must be of the form $u_\tau f(u_\tau y/\nu)$ where $u_\tau \equiv \sqrt{(\tau_w/\varrho)}$ is called the friction velocity. From the discussion in previous chapters of the viscosity-independence of the energy-containing turbulent eddies, we can expect that the influence of viscosity will be confined to a very thin layer close to the wall.

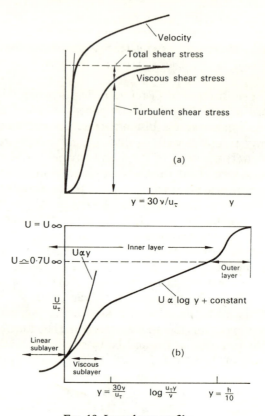

FIG. 18. Inner layer profiles.

The region where viscosity affects the shear stress and the energy-containing eddies is interesting, both in its own right and because the eddies found there are necessarily those best able to survive in the presence of strong dissipative forces—that is, they are probably more highly organized (simpler shapes) than the eddies elsewhere in the flow, and are therefore likely to be easier to understand. This seems to be confirmed by the flow visualization work of Kline and his collaborators (ref. 8 and later papers), Corino and Brodkey[9] and other authors.

Outside this layer, the shear stress (still equal to τ_w in Couette flow) will be produced entirely by the turbulent eddies so that $-\overline{uv}$ will be equal to u_τ^2, u_τ will be a typical measure of the *turbulent* velocity fluctuations, and $u_\tau y/\nu$ will be a turbulence Reynolds number based on a typical velocity fluctuation and a typical length scale: viscous effects on the energy-containing eddies ought to be negligible if this Reynolds number is sufficiently high. In practice, we find that if $u_\tau y/\nu$ is greater than about 30, the direct viscous contribution to shear stress (Fig. 18a) and the effects of viscosity on the shear-stress-producing eddies, but not of course the viscous dissipation in the smallest eddies, are small. Therefore for $u_\tau y/\nu > 30$ (outside the "viscous sublayer") we can eliminate the viscosity from the list of important variables, so that dimensional analysis shows that the mean velocity gradient is given by $\partial U/\partial y = u_\tau/Ky$. Note that the mean velocity is given by integrating $\partial U/\partial y$ with respect to y from the surface to the point considered and therefore the effect of viscosity on the flow very close to the wall still appears in the velocity (as opposed to the velocity gradient) at any point in the flow. According to this simple analysis, K is an absolute constant whose value is found experimentally to be about 0·41. We can see immediately that the ratio of turbulent shear stress to total (viscous plus turbulent) shear stress $\tau_w \equiv \varrho u_\tau^2$ is

$$(\tau_w - \text{viscous stress})/\tau_w = 1 - \frac{\mu}{\varrho u_\tau^2}\frac{\partial U}{\partial y} = 1 - \frac{\nu}{Ku_\tau y}$$

wherever the above formula for $\partial U/\partial y$ is valid, and this confirms the statement above that viscous stresses are small for $u_\tau y/\nu$ greater than 30. Since u_τ and K are both constants, we can integrate the formula for $\partial U/\partial y$ immediately to give

$$\frac{U}{u_\tau} = \frac{1}{K}\log y + \text{const}$$

for $u_\tau y/\nu > 30$.

Remembering the above dimensional analysis, we can see that this formula becomes

$$\frac{U}{u_\tau} = \frac{1}{K}(\log u_\tau y/\nu + A)$$

where A is another constant, about 2·05 for a smooth solid surface (Fig. 18b). If the surface is rough we find that, at distances from the surface large compared with the roughness height k, the formula for

$\partial U/\partial y$ still applies but the additive constant A is now a function of the "roughness Reynolds number", which dimensional considerations give as $u_\tau k/\nu$, and of the geometry of the roughness.

An alternative derivation of the formula for $\partial U/\partial y$ in the region fairly close to the wall is based on the turbulent energy equation. We again suppose that the typical scales of the energy-containing turbulence are u_τ and y, so that the triple products and pressure-velocity mean products appearing in the turbulent energy diffusion [(2) in Section 2.2] are proportional to u_τ^3, and the energy dissipation rate is proportional to u_τ^3/y. Now if y is small enough for du_τ/dx to be much less than u_τ/y, then both the mean advection and turbulent diffusion of turbulent energy in the x direction will be small compared to the dissipation rate, and since du_τ/dy is zero the turbulent energy diffusion in the y direction is zero. Therefore the turbulent energy equation reduces to "production" = "dissipation" or

$$\frac{\tau}{\varrho} \frac{\partial U}{\partial y} = c \frac{u_\tau^3}{y}$$

where c is the constant of proportionality in the expression for the dissipation. This immediately gives

$$\frac{\partial U}{\partial y} = c \frac{u_\tau}{y} \quad \text{or} \quad \tau/(\partial U/\partial y) = \frac{1}{c} \varrho u_\tau y$$

agreeing with the previous expression if we equate c with $1/K$. This derivation does not add much to the previous one except *a posteriori* support for the assumption that u_τ and y are sufficient scales for the energy-containing turbulence, but in cases where we are not so sure about the validity of local-equilibrium assumptions we have to start by considering the turbulent energy equation: if the transport terms seem likely to be small then we may hope that the "eddy viscosity", defined as $\tau/(\partial U/\partial y)$, will be simply proportional to the product of the velocity and length scales of the turbulence.

Since the energy dissipation ε is u_τ^3/Ky, the ratio of the length scale of the energy-containing eddies, y, to the Kolmogorov length scale $(\nu^3/\varepsilon)^{\frac{1}{4}}$, is $(u_\tau y/\nu)^{\frac{3}{4}}/K^{\frac{1}{4}}$, which is about 16 at $u_\tau y/\nu = 30$: this ties up with the minimal factor of 10 suggested in Section 2.5 for the separation of the energy-containing and dissipating ranges. The factor of 100 or more required for the existence of an inertial sub-range is not reached until $u_\tau y/\nu$ is greater than 300.

Now, clearly, the analysis above, which assumes that the eddy length scale is directly proportional to the distance from the surface, cannot possibly apply for y approaching h in the Couette flow. In fact, application of the analysis to *both* walls shows that it cannot possibly apply for y approaching $h/2$. In practice, it applies only to about $y = h/10$ (or the equivalent in other types of shear flow) although the profile may be apparently logarithmic for rather larger distances from the surface, evidently as the result of some two opposing effects. The distance from the surface at which this "inner law" or "law of the Wall" analysis breaks down depends upon the particular flow considered, and we will go on to consider some of the more common shear flows, which have additional complicating features.

3.4. Two-dimensional Boundary Layers[5,10,11] (Figs. 19, 20)

For simplicity, we continue to consider flows in which the W component of mean velocity is everywhere zero, remembering that the w component of fluctuating velocity is emphatically *not* zero. A boundary

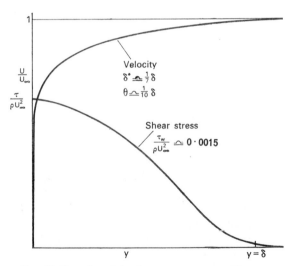

FIG. 19. Boundary layer in zero pressure gradient.

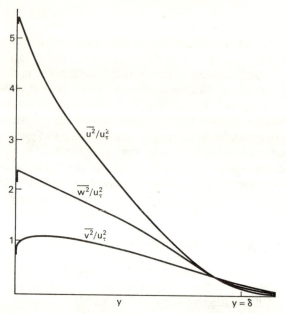

Fig. 20. Sketch of turbulent intensities in constant pressure boundary layer.
$u_\tau = \sqrt{(\tau_w/\varrho)}$, typically about $0.04U_\infty$.

layer is the growing shear layer between a solid surface and an infinite
stream: its essential property is that its thickness in the y direction,
δ say, is, except at very low Reynolds number, very small compared to
the length of the solid body. The mathematical use of this property
is called the boundary layer approximation, and since it also applies
to other types of thin shear layers (e.g. the Couette flow in the preceding
section) all thin shear layers are sometimes referred to collectively as
boundary layer flows: in this book, however, we use the term only in
its more precise meaning. In any thin shear layer, $\partial U/\partial x$ and $\partial V/\partial x$ are
very small compared with $\partial U/\partial y$.

At the outer edge of a turbulent boundary layer, the flow is "intermittent" (see Section 2.10) and the bounding surface between the turbulence and the irrotational flow is highly irregular. Since vorticity can
be transferred to the initially irrotational fluid only by the action of
viscosity, this interface is sometimes called the "viscous super-layer" by

analogy with the "viscous sub-layer" discussed above. In essence, the viscous super-layer is the exposed surface of the dissipating eddies near the edge of the flow and its thickness is expected to be of the order of the Kolmogorov length scale just inside the turbulent flow, but as long as the super-layer is thin compared to the dimension of the typical energy-containing eddies, it merely acts as an interface between the turbulence and the irrotational fluid, and the rate at which it propagates into the irrotational fluid, like the rate at which dissipation occurs in the very small eddies, is controlled by external influences and not by the viscosity. The overall shape of the interface is determined by the large eddies in the flow and it is these that specify the rate at which the irrotational flow is entrained by the turbulence. The rate of entrainment of mass into the turbulent flow also implies a rate of entrainment of momentum; that is, the large eddies contribute to the shear stress and also to the turbulent energy transfer (in fact at the outer edge, the increase of turbulent energy along a mean streamline is supplied almost entirely by diffusion from nearer the surface and local production and local dissipation of turbulent energy are both small by comparison). These large eddies seem to extend through almost the whole thickness of a well-developed shear layer and their behaviour determines the asymptotic shape of the space correlations. The "average" eddy seems to have a simple shape although there is considerable discussion about what this shape is. In any case, it is only an ensemble average (see Glossary) and the shape of individual eddies is highly irregular. Turbulence processes are quite weak; the v component in a turbulent boundary layer is never more than about 5 per cent of the free stream velocity, which means that disturbances introduced into the flow, or arising naturally, will spread out in the xy-plane at angles no more than 3 degrees to the x direction. Indeed, the angle between the edge of the boundary layer and the surface is very roughly of the order of 1 degree. Near the surface, in the region that is not directly affected by the behaviour of the turbulent flow in the outer part of the shear layer, the turbulence is statistically the same as in Couette flow—at least to an engineering approximation (the constants K and A seem to vary much more from experiment to experiment than from flow to flow!).

The average position of the edge of a boundary layer is difficult to define; the mean velocity tends asymptotically to the free stream value, as in a laminar boundary layer. For engineering purposes, it is usual to

define the boundary layer thickness as the distance from the surface
at which the mean velocity rises to some specified fraction of the free
stream velocity U_∞, frequently 0·995. The boundary layer thickness is
given the symbol δ. The quantities that appear in mass and momentum
balances for the whole boundary layer are the displacement thickness
δ^*, which is the distance by which the external streamlines are displaced
normal to the surface by the presence of the retarded flow in a boundary
layer, and the momentum-deficit thickness, or momentum thickness, θ,
which is the thickness of a layer of the free stream whose momentum
is equal to the amount by which the momentum of the flow is reduced
by the presence of the retarded flow in the boundary layer. The momentum thickness is a measure of the drag of the solid body on which the
boundary layer has developed and so it is this thickness that is most
frequently used in specifying the overall Reynolds number of the boundary layer. Very roughly, it is about one-tenth of the boundary layer
thickness δ. The latter is clearly the more suitable measure of eddy size.

For y greater than about 0·2δ (the "outer layer") the turbulence is
affected by the presence of the edge of the boundary layer and typical
eddy sizes are no longer directly proportional to distance from the surface.
Remembering that the rate of propagation of disturbances across the
boundary layer is quite slow, we can see that the quantity really affecting
the eddies for $y/\delta > 0\cdot2$ is not the local thickness δ but the thickness
perhaps 20δ upstream. This is one way of demonstrating that a turbulent
flow has a rather long "memory" of its previous history (see Sections 1.5
and 2.2): another way is to form a "time constant" by dividing the
turbulent energy per unit volume by the rate of production of turbulent
energy per unit volume. Both the turbulent shear stress and the turbulent
energy decrease to zero at the outer edge of the flow but the ratio of
the turbulent energy $\frac{1}{2}\varrho(\overline{u^2} + \overline{v^2} + \overline{w^2}) \equiv \frac{1}{2}\varrho\overline{q^2}$ to the shear stress $-\varrho\overline{uv}$
is about 3 over most of the flow so that

$$\frac{\text{energy}}{\text{production rate}} = \frac{\frac{1}{2}\varrho\overline{q^2}}{-\varrho\overline{uv}\,\partial U/\partial y} \simeq \frac{3}{\partial U/\partial y},$$

which is of the order of 10δ/U_∞ in the outer layer. This quantity may
be regarded as a "time constant" for the flow, analogous to that used
in discussions of first-order linear differential equations, although we
must, of course, recognize that the equations that govern turbulent

shear flows are neither linear nor of first order. The total duration of a phenomenon is of the order of three times its time constant which means that the lifetime of the larger eddies in a boundary layer will be of the order of $30\delta/U_\infty$, or a downstream distance of 30δ. We could make the idea of an eddy lifetime more precise by defining it as the distance r_1 at which the space-time correlation

$$\overline{u(\mathbf{x}, t)\, u(\mathbf{x} + r_1, t + r_1/U)}\,/\,\overline{u^2(\mathbf{x}, t)}$$

(that is, the space-time correlation following the motion of the fluid) fell to some small specified value, but the idea of an *eddy* lifetime is not to be taken too literally. If we were to mark an elementary square or cube in the flow at one instant, the fluid contained in it would be torn apart, tangled and spread over most of the boundary layer, so that any useful definition of lifetime would have to be made for a quantity averaged over the whole thickness of the boundary layer. However, the inner layer has such a short memory [$3U/(\partial U/\partial y)$ is typically about 3δ if $y = \delta/10$] that it can be treated as a local-equilibrium region matched to the outer flow, as in Section 3.4.

We can define two typical length scales for the boundary layer as a whole, the first being the boundary layer thickness δ and the second the ratio ν/u_τ, which is proportional to the distance from the surface at which the logarithmic velocity extrapolates to zero (this is about $0\cdot14\nu/u_\tau$). The latter length scale expresses the effect of viscosity in determining the inner boundary condition on the turbulent flow. The presence of these two length scales, from which, of course, many others can be derived, indicates that, even if the free stream velocity U_∞ is constant, we cannot expect the velocity profiles in the boundary layer to be exactly similar in the form

$$\frac{U}{U_\infty} = f\left(\frac{y}{\delta}\right)$$

found in laminar flow, since the additional parameter $u_\tau\delta/\nu$ will appear (in Couette flow, pipe flow and in certain other special cases, the Reynolds number based on u_τ and the flow thickness is constant and the velocity profiles *are* similar at different downstream distances). However, since the only effect of viscosity on the energy-containing eddies is applied by the boundary condition at the wall, we might expect that the velocity profiles measured by an observer moving with the free stream would be similar outside the viscous sublayer. From the point of view of this observer, the only effect of the wall on the flow is to transmit a shear stress τ_w, so we expect the velocity defect $U_\infty - U$ outside the viscous sub-layer to depend on τ_w (or u_τ), y and δ but not directly on v and U_∞ if the latter is constant, and dimensional analysis shows us that

$$\frac{U_\infty - U}{u_\tau} = f_1\left(\frac{y}{\delta}\right), \quad \frac{\tau}{\tau_w} = f_2\left(\frac{y}{\delta}\right), \quad \frac{\overline{\varrho u^2}}{\tau_w} = f_3\left(\frac{y}{\delta}\right),$$

etc. (Fig. 20). Strictly, we ought to include the rate of growth of the boundary layer, $d\delta/dx$, as some sort of measure of the history effect discussed in the last paragraph, but since the rate of boundary layer growth is a unique function of the Reynolds number if U_∞ is constant we cannot distinguish the growth effect from the true Reynolds number effect. However, both seem to be negligible for Reynolds numbers $U_\infty\theta/v$ greater than about 5000. The above equation for $U_\infty - U$ is called the "defect law" or "outer law" and it is expected to be valid all the way from the outer edge of the viscous sub-layer to the edge of the boundary layer itself. The negligible influence of $d\delta/dx$ on the defect law does not imply that the history effect is negligible but merely that it is, in some dimensionless sense, constant down the length of the boundary layer. For instance, we expect the ratio of the rate of change of turbulent energy along a streamline to the rate of production of turbulent energy at a given value of y/δ to be virtually independent of distance downstream, so that such quantities need not appear explicitly in any dimensional analysis.

The velocity defect law, above, is expected to be valid down to the edge of the viscous sub-layer, $y \simeq 30v/u_\tau$, so that it should overlap the logarithmic law in the region $30v/u_\tau < y < 0.2\delta$. If the two laws are simultaneously valid the defect law must have the same slope as the logarithmic law in this region and must therefore take the form

$$\frac{U_\infty - U}{u_\tau} = f\left(\frac{y}{\delta}\right) = -\frac{1}{K}\left[\log\frac{y}{\delta} + c\right].$$

Now if we substitute for U/u_τ from the logarithmic law we get

$$\frac{U_\infty}{u_\tau} = \frac{1}{K}\left[\log u_\tau\delta/v + A - c\right]$$

which is a relation between u_τ/U_∞ and $u_\tau\delta/\nu$ or between

$$c_f \equiv \tau_w/\tfrac{1}{2}\varrho U_\infty^2 = 2(u_\tau/U_\infty)^2$$

and the Reynolds number $U_\infty\delta/\nu$. K, A and c are absolute constants, determined by experiment. We have obtained the analytical form of this "skin friction" formula for a turbulent boundary layer by a refinement of dimensional analysis, without using any specific theoretical ideas about the turbulence itself. This being so, it is scarcely surprising that the same results can be obtained even when the dimensional analysis is disguised by an erroneous theory of turbulence, as long as the theory is dimensionally correct. For instance, the "mixing length" analogy between turbulence and molecular motion leads to the logarithmic law (although not to a numerical value for K), and the "eddy viscosity" in the outer layer of a constant pressure boundary layer is

$$\frac{\tau}{\partial U/\partial y} = \frac{\tau_w\delta}{u_\tau}\frac{f_2(y/\delta)}{f_1'(y/\delta)} = \varrho u_\tau\delta F(y/\delta) = \text{const } \varrho U_\infty\delta^* F(y/\delta)$$

from the above defect-law analysis, in agreement with a false analogy between laminar and turbulent flow which ignores the history effect.

This false impression of simplicity is soon dispelled when we consider more practical cases like that of the boundary layer on an aerofoil[10]. In order to produce lift, an aerofoil is shaped so that the pressure on the top surface is less than the pressure on the bottom surface, the pressure itself varying continuously over the perimeter of the wing section. This pressure gradient is, of course, linked by Bernoulli's equation to the variation of the velocity outside the boundary layer, and the boundary layer itself is directly affected by the pressure gradient, leading to departures from similarity even of the velocity defect profiles, leading in turn to changes in the shear stress and other properties of the turbulence. It is possible to distinguish special cases, "self-preserving" boundary layers, in which the ratio of a typical pressure gradient force to a typical wall stress force is independent of x, and in these cases similar velocity defect profiles and shear stress profiles are found, the exact similarity functions depending upon the ratio of pressure gradient to shear stress. Exactly self-preserving layers are not likely to be found in practice but they have been useful in the past for experimental and theoretical studies, being simpler to

deal with than the case of arbitrary pressure gradient. Since the turbulence is not directly affected by the mean pressure gradient, we expect
its structure to be relatively independent of the details of the pressure
distribution over the aerofoil or other body on which the boundary
layer is found, and most of the basic experimental work has been done
on the simplest case of a constant pressure ("flat plate") boundary
layer. This means, of course, that any changes in the turbulence structure that do occur in boundary layers in arbitrary pressure gradients
are not very well documented. As in the case of a laminar boundary
layer, too rapid a retardation of the free stream causes separation
(Fig. 26) although, because turbulent shear stresses are very much
larger than viscous shear stresses, a turbulent boundary layer can with-
stand a very much stronger pressure gradient than a laminar boundary
layer.

The mechanism by which the large eddies in the flow appear and cause
entrainment of non-turbulent fluid is not well understood. The rate of
production of turbulent energy per unit volume is a maximum within
the viscous sub-layer [in fact $-\varrho\overline{uv}(\partial U/\partial y)$ is a maximum when
$-\varrho\overline{uv} = \tau_w - \mu(\partial U/\partial y) = \tau_w/2$, which occurs at about $u_\tau y/\nu = 12$] but
the turbulent eddies produced in this region are several orders of mag-
nitude smaller than the boundary-layer thickness and it would take many
times their original lifetime for them to evolve into the large eddies
discussed above, so that we could not really regard them as the same
eddies. In fact, it is clear that eddies produced at $u_\tau y/\nu \approx 12$ must have
forgotten the circumstances of their birth by the time they reach
$u_\tau y/\nu = 30$ because the effects of viscosity are negligible for $u_\tau y/\nu$ greater
than 30. Also, since a typical eddy size increases directly proportional
to distance from the wall, the number of eddies per unit area in the
xz-plane decreases as $1/y^2$ and only a very small percentage of the eddies
starting at $u_\tau y/\nu = 12$ could retain their identity out to $y/\delta = 0.2$. It
seems more likely that the large eddies in the outer layer of the boundary
layer arise from the region near $y/\delta = 0.2$ where the inner layer eddies
themselves are not too small compared with the boundary layer thickness,
but there is a continuing temptation in the analysis of turbulence to
try to draw sharp distinctions between processes that are really hazily
defined and, in this elementary account, I must resist this temptation.
Although we have quite a large amount of documentary evidence on
the behaviour of the turbulent intensities, scales and other quantities

in a boundary layer, it has not yet been combined into a definitive explanation of the detailed physics of the flow. The best introduction is the review article by Rotta[12]: see also the recent references on p. 209.

3.5. Three-dimensional Boundary Layers

In most practical cases, the W component of mean velocity will not be zero. There will be a velocity gradient $\partial W/\partial y$ and the component of Reynolds stress in the yz-plane, $-\varrho\overline{wv}$, analogous to $\varrho\overline{uv}$ in the xy-plane, will not be zero. Three-dimensional shear layers over a solid body divide themselves into two classes according to whether $\partial U/\partial z$ is or is not small compared to $\partial U/\partial y$. Over the main part of, say, a swept-back wing, the boundary layer is quite thin compared to the length scale of any changes in the spanwise (z) or chordwise (x) directions and so $\partial U/\partial z$ is much less than $\partial U/\partial y$, and similarly $\partial W/\partial z$ is much less than $\partial W/\partial y$. However, close to the junction between the wing and body of an aeroplane or in the corner of a duct of non-circular cross-section, velocity gradients in both the y and z directions become important. These two classes of three-dimensional boundary layer have been called "boundary sheets" and "boundary regions" respectively: boundary regions are most conveniently discussed in connection with duct flows, in the next section. The turbulence structure of boundary sheets will not differ very much from that of two-dimensional boundary layers because the turbulence itself is *always* three-dimensional and small changes in the direction of the shear stress vector are not expected to have very much effect on properties like the shape of the frequency spectra. However, the need to calculate the direction of the shear stress as well as its magnitude introduces difficulties, because we cannot necessarily assume that the direction of the shear stress is the same as the direction of the velocity gradient, i.e.

$$\frac{-\varrho\overline{uv}}{-\varrho\overline{wv}} = \frac{\partial U/\partial y}{\partial W/\partial y}.$$

This relation would hold for viscous shear stresses in a laminar boundary layer, but in turbulent flow the processes that determine the velocity gradient and the processes that determine the shear stress are not closely connected (the turbulent energy equation shows that it is only the rate of change of turbulent intensity and not the turbulent intensity itself that depends upon the local velocity gradient, and exactly the same applies to the shear stress): therefore, the angle between the shear stress and the velocity gradient depends on the history of the flow. The angle is, however, not usually very large and has even been assumed zero in some calculation methods, apparently without violently contradicting the rather small amount of available experimental evidence. It is fair to say that if we understood two-dimensional shear layers better, we would be able to calculate three-dimensional ones without any extra work. However, in the present state of knowledge, experimental studies of three-dimensional flows are urgently needed for fundamental purposes and for the development of prediction methods.

3.6. Duct Flows[38] (Fig. 21)

At the entrance to a duct, boundary layers form on each wall and eventually meet and interact near the centre line. For simplicity let us consider the case of a duct with a rectangular cross-section of high aspect ratio, sometimes called a two-dimensional channel, although the term "channel" should properly be restricted to a flow with a free surface. The interaction takes place between two two-dimensional

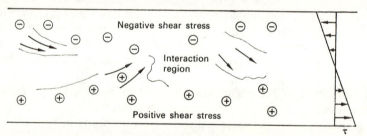

FIG. 21. Interaction between two sides of duct flow.

boundary layers with opposite signs of shear stress and velocity gradient. This interaction of two shear layers with an angle of 180° between the shear stresses is potentially much more difficult than the problem of the three-dimensional boundary layer with perhaps only 10° or 20° change of shear-stress direction across the flow, although if we understood the one, we would be in a better position to understand the other. In a symmetrical duct, the shear stress and mean velocity gradient are both zero on the centre line but this disguises the fact that the flow on the centre line is actually the time mean of two opposing flows, as shown in Fig. 21. The interface drawn between the two flows may seem to have some connection with the interface at the edge of a single turbulent boundary layer but, except in the very early stages of the interaction in the entry region, this is not a particularly useful concept. We note that if a turbulent eddy bearing a positive shear stress erupts into a region of negative velocity gradient, the production of turbulent energy will be negative and the eddy will be attenuated rapidly. In asymmetrical duct flows (for instance, the case of an annular duct flow like that shown in cross-section in Fig. 22) the situation in the region of interaction between the two shear layers is more obviously complicated—for in-

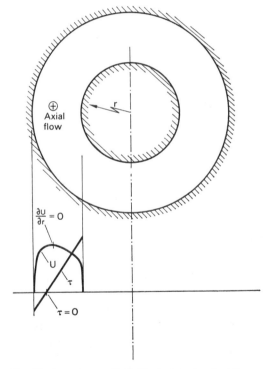

FIG. 22. Asymmetry of profiles in annular duct flow.

stance the position where the shear stress changes sign may not be the same as the position where the velocity gradient changes sign. This is, of course, perfectly explicable in terms of the discussion above and the general principle of loose connection between turbulent quantities and mean flow quantities but, in this particular case, a region of *negative* production of turbulent energy appears and has attracted some interest. In reality, it is not a new physical phenomenon: in no sense do the turbulent vortex lines straighten themselves out and disappear into the mean flow; it simply happens that, in this region, the turbulent eddies transfer *some* of their energy to the mean flow instead of the reverse, as is more usual. The energy is supplied by diffusion from elsewhere and the turbulent kinetic energy equation is still valid.

The circular pipe is a more difficult case than the duct of high aspect ratio because $-\varrho \overline{uv}$ on the centre line is the (zero) sum of turbulence from all parts of the perimeter with shear stress in all possible planes. The flow looks innocuous but is in reality diabolically complicated. Closer to the walls, the flow in ducts and pipes obeys the same universal laws as the flow in Couette flow or in a boundary layer, and velocity defect laws, different for pipe and duct, are also obeyed.

The flow in a duct does not become fully developed (independent of downstream distance) until fifty or more duct heights from the entrance: also, disturbances introduced into the flow by bends, branches and changes of section may take an equally long time to decay. This means that there is a good deal of engineering interest in perturbed or non-fully developed duct flows. From the fundamental point of view, such flows further illustrate the dependence of turbulent flow on upstream history already noted in Section 3.4. Of course, a laminar flow depends on upstream history as well, but in that case the shear stress adjusts at the same rate as the velocity gradient. A curious effect due to slow adjustment of the shear stress is the behaviour of the mean velocity profile in a developing duct flow. Near the entry, the velocity profile is flat-topped, with thin boundary layers: after the boundary layers meet, the profile becomes moderately peaked in the centre, but the peak gradually flattens out again to give a very shallow maximum in the velocity profile in fully developed flow. The explanation is that, when the boundary layers meet, the shear stress profile in one-half of the duct is roughly like that shown for a boundary layer in Fig. 19 (with $y = \delta$ at the centre of the duct) and so the shear stress gradient over most of the central region is greater than if the shear stress varied linearly from the wall to the centre as it does in fully developed duct flow: consequently, the flow in the central region is accelerated.

In the corners of ducts, or in the corner between a wing and a body (the case of a boundary region mentioned above), the two shear stress fields that interact are at 90° to one another[13]; in a square duct the flow is, of course, statistically symmetrical about the bisector of the corner but this still permits the existence of weak mean flows in the yz-plane, called secondary flows, which normally take the form of longitudinal vortices (see Fig. 13). An equation for the x component of mean vorticity can be obtained from the Reynolds equations (the time-average Navier–

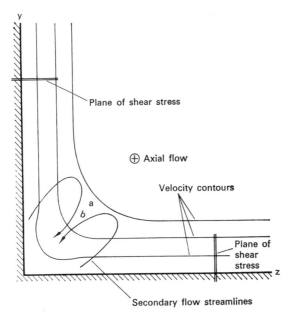

FIG. 23. Turbulent flow in a streamwise corner.

Stokes equations) by differentiating the W component equation with respect to y and subtracting the V component equation differentiated with respect to z, giving an equation for $\partial W/\partial y - \partial V/\partial z$. It appears from this equation and from physical considerations that the secondary flows are driven as much by Reynolds normal stress gradients as by Reynolds shear stress gradients. It can be seen, for instance, that the gradients of $\overline{v^2}$ and $\overline{w^2}$ between points a and b in Fig. 23 are in the right sense to maintain the secondary flow circulation shown. It must be remembered that these Reynolds stress gradients also produce pressure gradients: in the case of a simple two-dimensional shear layer, the pressure gradient normal to the surface is equal to the gradient of $-\varrho\overline{v^2}$ normal to the surface, assuming that the V component of mean velocity is negligible, but in corner flows the Reynolds stress gradient is not exactly equal to the pressure gradient and, in fact, it is the difference of the two that produces the secondary flow, although the pressure gradients do not appear in the vorticity equations. We see that to calculate this type of flow we have not only to understand the effect of the interaction of two shear stress fields, but also to calculate the normal-stress gradients occurring in the flow.

The above considerations apply to straight ducts. If the duct is curved, a longitudinal component of mean vorticity can arise even in an inviscid flow: it can be seen from the vorticity equation that if a flow with vorticity in the z direction is turned through an angle α in the xz-plane, a component of streamwise vorticity equal to 2α times the original z-component vorticity will appear. In the junction between a wing and a body or between a turbo-machine blade and the casing, both sources of vorticity will be important. The sense of the vorticity generated by the quasi-inviscid motion of a shear layer round an obstacle can be inferred by supposing that the vortex lines in the original flow move with the fluid and are wrapped round the obstacle. The flow in the main part of the corner region is complicated enough but the flow near the leading edge is more complicated still and consideration of it will be postponed until Section 3.8.

3.7. Jets, Wakes and Plumes (Figs. 24, 25)

"Free" turbulent flows have a change of sign of shear stress and velocity gradient on the centre line, like duct flows, but also a free edge, like a boundary layer, and their properties can be inferred qualitatively from the properties of these flows. The ratio of a typical turbulent velocity fluctuation to the local mean velocity reaches high values near the edges of jets and plumes in still air (the ratio does not reach infinity because the entrainment of fluid by the turbulence induces a velocity directed towards the centre line, but reliable measurements can be obtained with a hot-wire anemometer only if an axial velocity, sufficient

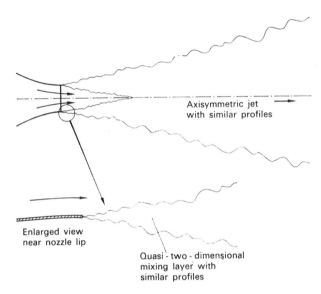

Axisymmetric jet
with similar profiles

Enlarged view
near nozzle lip

Quasi - two - dimensional
mixing layer with
similar profiles

FIG. 24. Regions of "similarity" in flow from a nozzle of arbitrary shape into still air.

to reduce the ratio to, say, 0·3, is superimposed on the flow). Even the ratio of the r.m.s. *u*-component intensity to the *centre-line* velocity can reach 0·2 or 0·3 near the point of inflexion of the velocity profile. These high values tend to counter the attractive simplicity of the boundary conditions, which would otherwise make jet flows an ideal subject for basic research: to the experimental engineer they imply difficulties in making and interpreting measurements and perhaps the need for special techniques.

Far from the orifice, the properties of a turbulent jet in still air (Fig. 24) are independent of the orifice shape and size, and depend only on the momentum flux, or thrust $M = \varrho U_0^2 A_0$ where U_0 is the (uniform) speed of flow through the orifice and A_0 is its area. In particular, dimensional analysis gives

$$U = \frac{1}{x} \sqrt{\left(\frac{M}{\varrho}\right) f_1\left(\frac{y}{x}\right)}$$

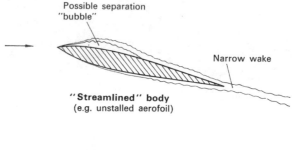

Possible separation "bubble"

Narrow wake

"Streamlined" body
(e.g. unstalled aerofoil)

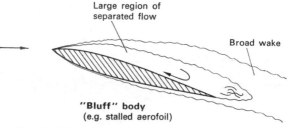

Large region of separated flow

Broad wake

"Bluff" body
(e.g. stalled aerofoil)

FIG. 25. Wakes of "streamlined" and "bluff" bodies.

for $x \gg \sqrt{A_0}$ where x is measured from an "effective origin" near the orifice, the precise origin depending on the details of the flow from the orifice. The velocity profile looks very much like the Gaussian curve in Fig. 9. Viscosity does not affect the mean flow or the energy-containing eddies except at very low Reynolds numbers or very near the orifice. If the orifice is a slit of height h and width $w \gg h$, then

$$U = \sqrt{\left(\frac{M}{\varrho x w}\right) f_2\left(\frac{y}{x}\right)}$$

in the region $h \ll x \ll w$: for $x \gg w$ the first formula applies again.

For $x < 4 - 5h$ in the case of a slit or $x < 4 - 5$ diameters in the case of a near-circular nozzle, the flow may be analysed with respect to an origin at the lip and dimensional analysis gives

$$U = U_0 f_3 \left(\frac{y}{x - x_0} \right)$$

where this time we have represented the effective origin, $x = x_0$, explicitly. This shear layer between a uniform stream and still air is called a "mixing layer" or "half-jet". The orifice boundary layer must be quite thin if similar flow is to be established before the shear layers from the different parts of the circumference meet. The shear layer into which a separated boundary layer, from an aerofoil or other body, develops is much more complicated because of the effects of streamline curvature, pressure gradients, reversed flow and the finite thickness of the separating boundary layer: however, empirical data from the ideal mixing layer between a uniform stream and still air have been used successfully in approximate calculation methods.

Far enough downstream of the body that produces it, a wake (Fig. 25) also becomes self-similar, with scales that depend only on the drag of the body and not on its shape. However, there are few practical cases where the behaviour far downstream is of interest: in the case of aerofoils or turbomachine blades the main need is to predict the wake for a fraction of a chord length downstream of the trailing edge, in the one case to estimate the profile drag and in the other to assess the effects on the next blade row. These problems are usually tackled by extensions of boundary-layer concepts. In the case of a bluff body, there is a region of reversed flow immediately downstream which must be treated on its own merits (see Section 3.9). The wakes of three-dimensional lifting bodies (aeroplanes) interact with the trailing vortices, whose rate of decay is of some interest because of their possible effects on following aircraft. The trailing vortices of delta wings are formed by the rolling up of vortex sheets which spring from the *leading* edge and greatly affect the flow over the wing. The rate at which the vortex sheets thicken and merge into vortex cores (Fig. 7) depends on turbulent mixing.

The wall jet is the flow obtained by blowing a jet parallel to a solid surface: its inner region behaves rather like a boundary layer and its outer region like a jet.

Plumes are closely similar to jets, but are driven by sources of buoyancy rather than momentum. Buoyancy per unit volume, $g(\varrho - \varrho_\infty)$, is closely related to enthalpy if the buoyancy is due solely to temperature changes rather than addition of foreign fluid, and obeys a conservation equation closely related to that for enthalpy (Appendix 1). In large-scale plumes in the atmosphere, the variation of the undisturbed density ϱ_∞ with height introduces complications in the equations. The conservation equation for momentum contains the buoyancy: neglecting density changes except in the buoyancy term (the "Boussinesq approximation") and assuming g and ϱ_∞ to be constant then, taking z as the vertical direction as usual in meteorology, we have

$$\frac{d}{dz} \iint \varrho w^2 \, dx \, dy = g \iint (\varrho - \varrho_\infty) \, dx \, dy.$$

The buoyancy conservation principle is

$$\frac{d}{dz} \iint (\varrho - \varrho_\infty) \, w \, dx \, dy = 0,$$

so that the behaviour of the momentum flux depends on the details of the velocity and density profiles.

The practical interest is usually in a plume in a crosswind, which, unlike the simple plume discussed above, will not usually exhibit similar profiles. Also, real-life plumes usually contain foreign fluid.

Buoyancy fluctuations can affect the turbulence directly if, as is usual, they are correlated with the velocity fluctuations: for instance, density fluctuations ϱ' contribute an extra rate of production of turbulent energy, $-g\overline{\varrho'w}$ in the present notation, which may greatly augment or completely damp the turbulence according to the sign of the correlation. The ratio of (minus) the buoyant production term to the ordinary shear production term is one of the possible definitions of the Richardson number. Buoyant flows tend to be stable (damped) if the Richardson number is positive and unstable (augmented) if it is negative.

3.8. Atmospheric and Oceanic Turbulence

Wind loads on earthbound structures are usually important only when the wind is strong. In this case the Richardson number is small and buoyancy forces unimportant, and the flow in the first few hundred metres above the ground is closely the same as in a laboratory boundary layer. The two remaining difficulties are that the "free stream" conditions are not well defined and that the surface roughness is large, inhomogeneous and difficult to specify.

Fluctuating loads on aircraft depend on the vertical component of wind velocity (which alters the incidence) as well as on the horizontal component. Therefore a moderately calm day with pronounced thermal activity may cause as much disturbance to a low-flying aircraft as a windy day (a "thermal" is a body of air rising from a transient source of buoyancy, as distinct from a "plume" from a continuous source).

At altitudes of more than roughly a thousand metres, the horizontal component of mean velocity is affected more by pressure gradients and Coriolis forces than by Reynolds stresses so that the upper atmosphere is not usually thought of as turbulent, except in jet streams and ascending cumulus clouds. What airline pilots call "turbulence" is often a train of gravity waves in a stably stratified region, but it is now clear that gravity waves can break down into turbulence due to Kelvin–Helmholtz instability, and some of our ideas about the upper atmosphere and the ocean are currently being revised. Gravity waves in a fluid whose density decreases continuously with increasing height behave in qualitatively the same way as waves in a fluid whose density decreases discontinuously at a free surface. Not only do they interact with pre-existing turbulence in a complicated way, but if a gravity wave "breaks" in a region of unusually strong density gradient, a patch of new turbulence can be generated. These patches gradually decay into weak background turbulence because the mean velocity gradient is too small for shear production of turbulence to outweigh the (negative) buoyant production due to the stable mean density gradient. However, the contribution of the turbulence generated in the patches to the overall rates of vertical transfer of mass, momentum and enthalpy may be significant. The sequence described above is self-perpetuating, because a region of unusually strong density gradient that arises in a weakly turbulent, stably stratified fluid is likely to damp out the turbulence in its vicinity so that enthalpy and momentum can be transferred vertically only by molecular diffusion down the density and velocity gradients, which therefore become even stronger until a passing gravity wave breaks and starts the process anew. It is not yet certain how common and important this process is in practice but it is undoubtedly a source of clear air turbulence of smaller length scale than the so-called turbulence of the gravity waves themselves.

The weather results from an interaction between the higher regions of the atmosphere, in which enthalpy and water vapour are transported over long distances, and the turbulent boundary layer which transfers enthalpy and water vapour directly to and from the surface. Short-range transport in the boundary layer is not an important factor in the weather, except in the case of local fogs, but it is extremely important in determining the dispersion of wanted or unwanted pollutants released near the ground. Clearly, the best way of reducing the concentration of unwanted pollutant is not to release it in the first place: the study of turbulence in combustion chambers and chemical reactors has a part to play in this, but all that a study of atmospheric turbulence can do is to predict the dispersion of a given quantity of pollutant and possibly suggest the least harmful place to release it.

Micrometeorology,[4] the study of the first few hundred metres of the atmosphere, combines the study of boundary layers and buoyancy effects. Experimental difficulties, connected with instrumentation as well as with the ill-defined nature of the boundary layer mentioned above, are considerable. Although fairly satisfactory correlations have been obtained for the effect of buoyancy on the dissipation rate and the apparent eddy viscosity and eddy conductivity, the scatter of the experimental points around these correlations is rather larger than would be accepted in laboratory experiments and rather too large to discriminate adequately between competing theories. The task of setting up a laboratory boundary layer with significant buoyancy effects but a reasonably small temperature difference is difficult, but some elegant experiments have been done.

We will give one example of the insertion of buoyancy effects into an isothermal-flow formula: for more details see ref. 4.

In Section 3.3 dimensional analysis was used to derive the formula

$$\frac{\partial U}{\partial y} = \frac{(-\overline{uv})^{\frac{1}{2}}}{Ky}$$

for the velocity gradient near (but not too near) a solid surface. Buoyancy effects introduce a new quantity

$$L = \frac{(-\overline{uv})^{\frac{3}{2}}/K}{\text{buoyant production}}$$

having the dimensions of length and called the Monin–Obukhov length. Dimensional analysis tells us that now

$$\frac{\partial U}{\partial y} = \frac{(-\overline{uv})^{\frac{1}{2}}}{Ky} f\left(\frac{y}{L}\right)$$

or, as a first approximation for small buoyancy effects,

$$\frac{\partial U}{\partial y} = \frac{(-\overline{uv})^{\frac{1}{2}}}{Ky}\left(1 + \beta\frac{y}{L}\right)$$

where β, like K, is an empirical constant. Use of the original formula for velocity gradient in the definition of Richardson number given at the end of Section 3.7 shows that to a first approximation y/L is equal to the Richardson number. We can now integrate to get

$$\frac{U}{u_\tau} = \frac{1}{K}(\log y + \text{const}) + \beta y$$

where $u_\tau = (-\overline{uv})^{\frac{1}{2}}$ as usual. Since micrometeorology is exclusively concerned with rough surfaces without a conventional viscous sublayer the most convenient form for the constant of integration gives

$$\frac{U}{u_\tau} = \frac{1}{K}\log y/z_0 + \beta y$$

where z_0 is called the roughness length (in meteorology z is used for the distance normal to the surface but here we have stuck to y to avoid confusion with Section 3.2).

Oceanic turbulence (oceanographers measure z downwards!) is qualitatively similar to atmospheric turbulence except that effects produced by the condensation of water vapour (i.e. clouds and rain) are absent. Buoyancy effects can be produced either by temperature differences or by salt concentration differences, and the two can sometimes oppose one another.

3.9. Separated Flows (Figs. 26, 27)

In a two-dimensional separated flow like that behind a long cylinder transverse to the stream, or in a spanwise obstacle or cavity on a surface, we can again neglect the z component mean velocity and the momentum

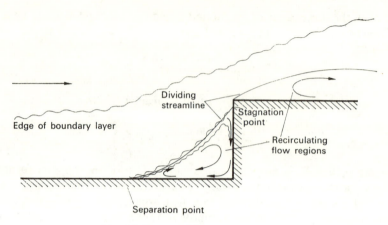

FIG. 26. Separated flow ahead of step.

transport in the z direction. However, one of the simplifying features of boundary layers is lost, namely the assumption that the shear layer is thin and nearly plane. Large changes may occur in the direction of the velocity, the rate of strain and the shear stress in the xy-plane, and the effect of normal stresses may be as large as that of shear stresses. Moreover, the pressure may vary considerably both in the x direction and in the y direction and two or more shear layers may interact. As in laminar flow, one of the mean streamlines is a boundary between the main part of the flow and a "dead water" region. Unfortunately, the latter term is a misnomer and quite large values of velocity and shear stress can occur. For instance, in experiments on the "dead water" region behind forward-pointing cones in a low-speed flow[14], it was found that a forward-moving jet emerged from the high-pressure region at the rear of the separated flow bubble with a velocity of about 0·4 of the free stream velocity. The effects in two-dimensional flow seem to be less spectacular but a jet with a velocity of about 0·25 of the free stream velocity is deflected downwards from the stagnation region near the top of a forward-facing step (Fig. 26) and seems to exert a considerable influence on the separated region. There are no useful calculation methods for turbulent separated flows, except for simple cases like the flow over a forward-facing or rearward-facing step which has few parameters and in which a single shear layer is still easily identifiable.

One of the subsidiary difficulties that occurs is that turbulence seems to be quite sensitive to the "centrifugal force" effects which occur in a curved shear layer and which seem to be roughly analogous to the effects of buoyancy in a gravitational field. It is known that a laminar flow is likely to be unstable if the angular momentum decreases outward from the centre of curvature of the streamlines: the mixing rate of a turbulent shear layer with angular momentum decreasing outward can be considerably augmented and, if the curvature is strong enough, the mean flow may develop a spanwise periodicity, caused by nearly-steady longitudinal vortices of the same sort that occur in the instability of a laminar flow on a curved surface. Similar effects occur in a rotating

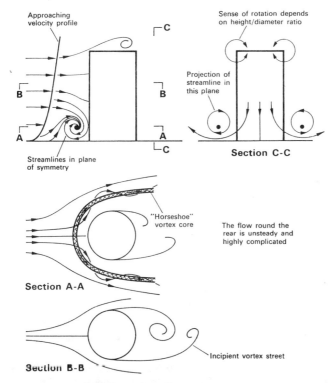

FIG. 27. Much-simplified sketch of flow past an obstacle in a boundary layer (building in a wind).

flow, where the apparent body force is the Coriolis force rather than the centrifugal force.

It is often found that separated turbulent flows are quite strongly unsteady, sometimes randomly and sometimes in a quasi-periodic sense leading to a vortex street in the wake. This is just a particularly severe example of the fact that separated flows interact with the main quasi-inviscid flow.

Three-dimensional separated flows (Fig. 27) are the most complicated form of turbulence, containing all the difficult features of two-dimensional separated flows plus the effects of stretching of *mean flow* vorticity. Taking as an example the flow of the atmospheric boundary layer round a tall building, we see that the elementary spanwise vortex lines that make up the mean velocity gradient will be wrapped round the building and stretched (thus increasing their vorticity) so that a horseshoe-shaped vortex is set up, with its core near ground level where the vorticity in the approaching boundary layer is strongest. Even in laminar flow, complicated flow patterns are set up: when the flow is turbulent and fluctuating vorticity is added to the mean vorticity, the detailed behaviour is indescribable, the solid surface being covered with subsidiary (unsteady) turbulent boundary layers moving in a variety of directions and separating when the pressure gradients become too large.

A rather simpler case is the flow round a blade root or a wing–body junction: the separated region ahead of the streamlined body is still present but the flow round the body itself is at least roughly in the direction of the main stream and the horseshoe vortex remains close to the streamwise corner. Generally speaking, the separated-flow problems for which accurate answers are required are those in which a fairly small separated region is imbedded in a fairly simple flow, but there is a great need for qualitative understanding of the more complicated flows.

Few turbulence measurements have been made in separated flows, whether two-dimensional or three-dimensional. The time is ripe for basic research work on the problem, both to help the industrial experimenter who has to face separated flows in practice and to provide data for the prediction methods that are now becoming mathematically feasible: in the absence of reliable measurements, such methods amount to little more than guesswork although some progress has been made in the simpler cases, like the flow over a step (Fig. 26) where a single thin shear layer is identifiable. Necessarily, the more complicated flows must be treated, either theoretically or experimentally, as combinations of the simpler flows described in Sections 3.1 to 3.7.

3.10. Heat and Mass Transfer

The turbulent heat transfer in the y direction, per unit area per unit time, is $Q \equiv \varrho c_p \overline{\theta v}$, where ϱ is the (constant) density, c_p the (constant) specific heat and θ the temperature fluctuation ($\overline{\theta} = 0$). Similarly, the turbulent transfer of concentration of a foreign fluid or the number-density of solid particles, c say, is $\overline{vc}$, if the fluid or solid particles move with the host fluid without affecting its motion. The *necessary* condition for this is that the density of the mixture shall be nearly the same as the density of the host fluid: the same condition is necessary *and* sufficient for heat transfer. Further obvious conditions for concentration are that no chemical reactions shall occur, and that the terminal velocity of a solid particle and the time required to reach it shall be much less than the velocity and time scales of the flow. These conditions are satisfied by most industrial pollutants.

Taking the case of heat transfer for simplicity we can find equations for the rate of change of $\overline{\theta^2}$ or $\overline{\theta v}$ along a mean streamline, analogous to the turbulent energy equation and the other equations mentioned in Section 2.2 (see Appendix 1). Unfortunately the temperature equations are appreciably different from the velocity equations (mainly because pressure fluctuations do not directly affect the temperature field) so we cannot expect a quantitative relation between $\overline{\theta v}$ and $\overline{uv}$ to hold in general. Indeed the only simple dimensionally correct relation not involving the further unknown $\overline{\theta^2}$ would be a relation between the turbulent diffusivities of heat and momentum

$$\frac{\overline{\theta v}}{\partial T/\partial y} \equiv \frac{Q/\varrho c_p}{\partial T/\partial y} = \frac{1}{\mathrm{Pr}_t} \frac{\overline{uv}}{\partial U/\partial y}$$

where Pr_t is evidently the turbulent analogue of the molecular Prandtl number $\mu c_p/k$, the ratio of the diffusivities of momentum and temperature: in meteorology the reciprocal of this ratio is used in discussion, with symbol K_H/K_M. However, we have seen that in general the turbulent momentum diffusivity does not obey simple rules, and the same goes for the heat diffusivity. Therefore the turbulent Prandtl number is not a constant or a simple function except in special cases. One such case is the flow near a heated surface, where the arguments used in Section 3.3 to derive the logarithmic velocity profile can be adapted to derive a logarithmic temperature profile. At distances from the surface small compared to the total width δ of the turbulent flow (say $y < \delta/10$) *and* small compared to the total width δ_T of the region of significant temperature variation (say $y < \delta_T/10$ where necessarily $\delta_T \le \delta$), we expect the temperature difference between the fluid and the wall, $T - T_w$, to depend only on (i) the heat transfer from the wall, Q_w, with dimensions of energy per unit area per unit time, (ii) the distance from the wall y, (iii) the properties of the fluid ϱ, c_p and k, and (iv) the velocity field, which in turn depends on $\tau_w \equiv \varrho u_\tau^2$, y, ϱ and ν. Dimensional analysis gives (omitting a kinetic-heating group unimportant at low speeds)

$$\frac{u_\tau(T - T_w)}{Q_w/\varrho c_p} = f\left(\frac{u_\tau y}{\nu}, \frac{\varrho c_p \nu}{k}\right)$$

where the second group on the right is of course the molecular Prandtl number. If neither viscosity nor thermal conductivity affect the temperature profile (which requires $u_\tau y/\nu > 30$ *and* $u_\tau y \varrho c_p/k > 30$) and if also $y < \delta_T/10$, the temperature gradient depends only on Q_w, u_τ and y and is therefore

$$\frac{\partial T}{\partial y} = \frac{-Q_w/\varrho c_p}{u_\tau K_\theta y}$$

so that

$$\frac{u_\tau(T - T_w)}{-Q_w/\varrho c_p} = \frac{1}{K_\theta}\left(\log \frac{u_\tau y}{\nu} + A_\theta\right)$$

inserting a negative sign because heat is transferred from high-temperature to low-temperature regions. $Q_w/(\varrho c_p u_\tau)$ has the dimensions of temperature and is called the "friction temperature" by analogy with the "friction velocity" $\sqrt{(\tau_w/\varrho)} \equiv u_\tau$. The dimensionless heat-transfer coefficient analogous to the skin-friction coefficient $\tau_w/\frac{1}{2}\varrho U_\infty^2$ is $Q_w/(\varrho c_p U_\infty(T_w - T_\infty))$, the Stanton number, abbreviated as St. K_θ, like K, is an absolute constant according to this simple analysis: it can be seen that the turbulent Prandtl number, as defined above, is K/K_θ. A_θ is a function of the molecular Prandtl number. Where the local-equilibrium analysis breaks down the turbulent Prandtl number depends on the history of the flow in an unknown way, although fairly adequate predictions of surface heat transfer in pipes and boundary layers can be made by assuming Pr_t = constant right across the flow: the temperature difference across the outer part of the flow, where Pr_t is not rigorously constant, is small compared to the temperature difference across the inner layer, so that the assumptions are not critical if only the Stanton number is required. If both the molecular and turbulent Prandtl numbers were unity, implying complete equivalence of heat and momentum transfer, then St would be equal to $\frac{1}{2}c_f$: in the more general case, $St/\frac{1}{2}c_f$, which is the reciprocal of a sort of overall effective Prandtl number, is called the Reynolds analogy factor. Various formulae have been proposed for the Reynolds analogy factor, as a function of Reynolds number and molecular Prandtl number, in the simple cases of pipe flow and the constant-pressure boundary layer. In these cases we can apply, to the outer part of the flow, the same arguments that led to the velocity defect law (Section 3.4)

$$\frac{U_\infty - U}{u_\tau} = f\left(\frac{y}{\delta}\right)$$

to obtain an outer layer temperature law

$$\frac{T - T_\infty}{Q_w/\varrho c_p u_\tau} = f\left(\frac{y}{\delta}\right)$$

where of course the two functions are different. Combining this with the logarithmic temperature law we obtain

$$\frac{T_w - T_\infty}{Q_w/\varrho c_p u_\tau} \equiv \frac{\sqrt{(c_f/2)}}{St} = \frac{1}{K_\theta}\left(\log\frac{u_\tau\delta}{\nu} + A_\theta - c_\theta\right)$$

where c_θ is an absolute constant. Using the logarithmic skin friction formula from Section 3.4 to eliminate $\log u_\tau\delta/\nu$ we get

$$\frac{St}{\frac{1}{2}c_f} = \frac{K_\theta/K}{1 + \sqrt{(c_f/2)}(A_\theta - A - c_\theta + c)/K} = \frac{K_\theta/K}{1 + \sqrt{(c_f/2)}f(Pr)}$$

where c_f is a function of Reynolds number. For a boundary layer in air, $Pr \simeq 0.7$, $f(Pr) = -1.09$. At very large Reynolds number, c_f tends to zero and $St/\frac{1}{2}c_f$ tends to K_θ/K, the reciprocal of the turbulent Prandtl number in the logarithmic layer. The reason why $St/\frac{1}{2}c_f$ is independent of molecular Prandtl number and outer layer conditions at high Reynolds number is that when c_f is small (i.e. u_τ/U_∞ small) the velocity difference across the viscous sublayer, say $12u_\tau$, and the velocity difference across the outer layer, say $5u_\tau$, are both small compared to U_∞—that is, the logarithmic part of the velocity profile covers nearly the whole range from zero to U_∞. Similar arguments apply to the temperature profile, so that the overall effective Prandtl number is the turbulent Prandtl number in the logarithmic layer.

At the corresponding point in Section 3.4, we went on to show that simple formulae derived by dimensional analysis in the constant-pressure boundary layer failed in boundary layers in pressure gradients, particularly adverse pressure gradients leading to separation. The same is in principle true for heat transfer but in practice the cases of greatest interest can usually be predicted to adequate accuracy by simple formulae or at most the invocation of a constant turbulent Prandtl number—given, of course, a good enough prediction of the velocity field. The two typical problems of heat transfer are to predict the heat transfer through the wall of a heat exchanger and hence the total length of heat exchanger required, and to verify that the maximum temperature of, say, a turbine blade does not exceed the creep limit. The accuracy required in the first case is of the order of 5–10 per cent, which is easy for simple shapes; the accuracy required in the second is better than 1 per cent (a few degrees in temperature difference) which is impossible, so that direct experiment is necessary. The problem intermediate between the easy and the impossible is to predict the heat transfer from heat exchangers of complicated shape to, say, 10 per cent accuracy. The alternative approaches are to rely on data correlations from systematic experiments or to attempt to calculate the velocity and temperature fields in detail, using existing knowledge of turbulence supplemented by a few experiments. The former approach deters innovation; the latter is less reliable for commonly used layouts. There is still plenty of work to be done in the study of turbulent heat transfer.

3.11 Turbulence in Non-Newtonian Fluids

Turbulence in Newtonian fluids is complicated enough but there is considerable practical interest in turbulence in non-Newtonian fluids, with stress proportional to some non-linear function of the rate of strain, or stress depending on both rate of strain and strain itself (viscoelastic substances). It is difficult to define a turbulent Reynolds number explicitly for such substances, but what really matters is the ratio of the length scale of the energy-containing eddies to the length scale of the largest eddies directly affected by the molecular properties. If this ratio is large, the energy-containing (and stress-producing) turbulence will be the same as in a Newtonian fluid, being independent of the precise mechanism of molecular dissipation (see the end of Section 1.6). However, the ratio is necessarily small near solid boundaries, where the length scale of the largest eddies is of the order of the distance from the boundary. Therefore non-Newtonian effects will certainly appear near solid boundaries, i.e. in the sub-layer, even if they are absent from the main part of the turbulent motion.

Clearly a substance in which stress increases indefinitely as the total strain increases is unlikely to become or remain turbulent, because turbulent diffusion causes very rapid extension of material lines (compare the extension of *vortex* lines). However, most viscoelastic substances have a low yield stress, beyond which the strain-dependent part of the stress ceases to rise: this is usually because the substance has a fibrous structure of some sort which breaks up if the strain becomes large. Examples are starch solutions and other long-chain polymer solutions: porridge, if stirred and then left to itself, will come to rest after a quasi-viscous motion and then rotate for a short time in the opposite direction as the elastic stresses relax.

Solutions of many complicated organic chain molecules have significantly less drag in turbulent pipe flow than the pure solvent, even for solution strengths of only a few parts per million which produce no appreciable effect in laminar flow. It is now clear that the cause is an increase in the additive constant A in the logarithmic velocity profile because of changes in the viscous sub-layer: what happens in the sub-layer is not understood in detail but it seems likely that the motion in the region closest to the wall where $\overline{uv} \simeq 0$ is, like purely laminar flow, almost unaffected, while the increase of turbulent intensity and shear stress with distance from the wall is inhibited by distortion of the molecules, leading to a larger velocity gradient for a given surface shear stress. A snag in this explanation is that individual molecules are far too small to behave like this: if tangling is responsible, hundreds or thousands of tangled molecules must be involved. The picture was obscured for some time because experimenters failed to realize that measuring instruments might also be affected by non-Newtonian phenomena: early measurements showed large changes in the *shape* of the logarithmic profile and other curious effects in the body of the flow. It is still not clear how far Pitot tubes and hot wires can be trusted: recent turbulence measurements[15] on a 0·01 per cent solution of a commercial polyacrylamide were made with a Doppler laser anemometer which will presumably be unaffected. It was found in this experiment that the *u*-component turbulence intensity $\sqrt{\overline{u^2}}/u_\tau$ near the edge of the linear sub-layer ($u_\tau y/\nu = 20$ in this case) was *higher* than in a Newtonian flow, by about the same factor as the sub-layer thickness. At first sight this contradicts the above explanation but it is known that the "turbulence" in the linear sub-layer consists largely of pulsations of the flow as a whole, driven by the turbulent eddies further out, so that one expects $\sqrt{\overline{u^2}}/U \approx$ constant, as found here.

If the fluid is strongly non-Newtonian (as in the case of stronger solutions of chain molecules) the effects may spread through the whole of the turbulent region: little progress has been made in this case.

CHAPTER 4

Measurement Techniques

"For the Snark's a peculiar creature, that won't
    Be caught in a commonplace way.
Do all that you know, and try all that you don't:
    Not a chance must be wasted today!"

THE hot-wire anemometer (by which will be understood, in this chapter, any "thermo-anemometer" relying on the rate of heat loss from an element whose temperature is deduced from its resistance[16]) is the instrument most widely used for fluctuation measurements. At the present time, the laser Doppler technique, in which instantaneous velocity is deduced from the Doppler frequency shift of light scattered from particles moving with the fluid, is being investigated by many groups of workers. Its advantage over the hot-wire method is that there is no solid sensing element in the flow—a double advantage because hot-wire probes and their supports may be large enough to disturb the flow, while the wire itself is small and temperamental. A good deal of development work is still needed to produce a laser Doppler system for everyday use, but commercial instruments are already available and can now compete with or complement the hot-wire anemometer in certain cases, particularly when smoke or dye can be easily introduced into the flow.

None of the other techniques for point measurements show any sign of rivalling the heated-element anemometer in general use, but I have included brief details of some that may be useful for special purposes. It seems likely that optical techniques may become more popular in the future: apart from point-measurement instruments like the laser Doppler anemometer, computer analysis of flow-visualization pictures is potentially a powerful method of collecting information about the turbulence

85

in a plane instead of merely at a point. The more specialized or speculative techniques are collected in a section at the end of the chapter, as a guide for workers with special problems and as an indication of possible future trends. Of course, advances in signal-processing techniques can be as helpful as advances in the sensing elements themselves: this subject is discussed in Chapter 6.

4.1. Hot Wires, Films and Thermistors (Fig. 32)

With modern advances in electronic circuitry and a general improvement in the design and construction of probes, the hot-wire anemometer can now be used for all the more common measurements by relatively inexperienced operators with no specialist electronic knowledge, although a specialist knowledge of turbulence is still needed to interpret the results. The sensing elements used are usually quite delicate mechanically and respond to changes in fluid temperature as well as changes of fluid velocity, so that frequent checks of probe calibration are essential. This point cannot be too strongly emphasized although attempts have been made in the past to minimize it. This means that absolute calibrations (for instance universal relations between a Nusselt number and a Reynolds number for a hot wire) are not usually very useful except as a rough check on calibrations. The simplest reason for this is that very fine wires cannot be made to an accurate diameter, nor can the diameter be easily measured to high accuracy, so that the uncertainty in wire Reynolds number is considerable: films, of course, have a very wide range of possible geometries. There are various other reasons, related to the differences between the idealized wires for which the absolute calibrations hold and the real wires used for anemometry, which make the individual calibration of new probes essential. If probe manufacture is carefully controlled, the difference between successive probes may be small enough for "calibration" to be reduced to one or two measurements to determine one or two constants in a theoretical or empirical relation between the speed of the flow and the voltage difference across the wire. Irrespective of this, and irrespective of the type of sensing element, sensitivity to temperature changes and the effects of dirt in the fluid stream make frequent recalibration desirable. The accuracy of the results obtained with hot-wire anemometers and other methods of turbulence measurement may also be degraded by improper choices of the processing equipment used to take statistical average values of the signals. The discussion in this book will be centred mainly on probe performance and processing equipment. Several reliable anemometer sets are available commercially and it is scarcely necessary to discuss the principles of their operation, except to give the new user enough information to enable him to choose between the different types of anemometer available. One point that should be made at the outset is that because of calibration drift the hot wire is not a very suitable instrument for mean velocity measurement.

4.2. Constant-current and Constant-temperature Operation (Fig. 28)

In constant-current operation a current through the wire is maintained constant and the wire voltage (resistance) is measured. Compensation

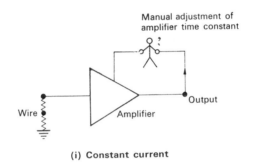

(i) **Constant current**

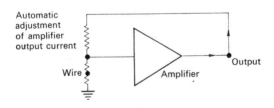

(II) Constant temperature

FIG. 28. Similarities between constant-current and constant-temperature operation.

for the thermal inertia of the wire, which attenuates its amplitude response to high frequency fluctuations by a factor $1/\sqrt{(1 + \omega^2 M^2)}$ where M is the "time constant" (see Section 5.7), is effected by feeding the hot-wire output through an amplifier whose gain rises with frequency as $\sqrt{(1 + \omega^2 M^2)}$ up to some limit set by the amplifier components so that the effective time constant ($\equiv 1/$bandwidth in radians/sec) is roughly the wire time constant divided by the "intrinsic" gain of the amplifier (i.e. the maximum gain attainable). Since the time constant depends somewhat on the operating conditions of the wire, the compensator gain has to be adjusted manually during the course of the experiment. Films do not have a simple time constant and are rarely operated at constant current. In the constant temperature system, the element temperature (resistance) is maintained constant by an amplifier in a feedback loop, usually a Wheatstone's bridge, which has the effect of compensating for the thermal inertia of the wire or film: again, the time constant of the whole system is roughly the wire time constant divided by the amplifier gain. The basic difference between constant-current and constant-temperature operation is, therefore, whether the compensation for thermal inertia is adjusted automatically or manually. Several points have been glossed over in this brief description and, in practice, there are considerable differences in the circuits. In the early days of electronics it was safer for compensation to be under manual control but modern equipment is so much more reliable that the ease of use of the constant-temperature system has made it more popular than the constant-current system. It has sometimes been asserted that the noise level of constant-temperature apparatus is inherently larger than that of constant-current apparatus because the feedback amplifier must, for stability reasons, have a wider frequency bandwidth than the desired effective bandwidth of the anemometer. However, if suitable filters are fitted to the apparatus, amplifier noise above the effective frequency limit can be cut off, and there is no other essential difference between the noise level characteristics of the two systems. The only advantage of constant-current equipment in this respect is that it is more straightforward to couple the wire to the amplifier by a transformer (which is ideally an amplifier producing no noise at all), thus increasing the signal to noise ratio. The constant-current transformer-coupled system is therefore sometimes preferred for measurements of very low r.m.s. turbulence intensities of the order of 0·1 per cent or less of the mean velocity: a fairly elaborate system is described in ref. 17. Recent developments in low-noise amplifiers with parallel input stages have largely superseded the transformer.

For very large turbulence intensities, the constant-temperature system is to be preferred because the time constant of a hot wire depends on the operating conditions or, for a given wire, on the speed of flow past the wire: therefore, large fluctuations in flow speed cause appreciable fluctuations in the time constant itself, with consequent errors in reproduction of high-frequency signals. This difficulty does not occur in the constant-temperature system because the "compensation" is controlled automatically by the amplifier: compensation for the modulation of time constant in the constant-current system would require a parametric amplifier whose high frequency gain was controlled by the wire voltage itself. It would not be impossibly difficult to devise such an amplifier but, in practice, the constant-temperature system is almost always preferred for high intensity turbulence measurements. It is still necessary to compensate for the distortion produced by the non-linearity of the wire calibration but this can easily be done by using a suitable "linearizer" (non-linear amplifier) such as a chain of biased diodes (see Section 6.1.3) fed directly from the anemometer and arranged so that the diode chain output is directly proportional to the mean velocity. If the r.m.s. turbulent intensity exceeds about 0·3 of the mean velocity, interpretation of wire readings becomes increasingly uncertain (Section 5.3).

In flows with large temperature fluctuations as well as velocity fluctuations (especially supersonic flows) it is necessary to operate the wire at several different temperatures (that is, several different ratios of velocity sensitivity to temperature sensitivity) so that the statistical properties of the velocity and temperature fluctuation fields can be deduced (see Chapter 7). It is somewhat easier to do this with a constant-current system than with a constant-temperature system because the overall frequency response and stability of the latter depend upon the wire temperature, partly because the time constant of the wire varies and partly because the amplifier gain depends on the output current. This is not a difficulty in normal use but if attempts are made to operate the wire at a temperature just above that of the fluid, the current will be small, the frequency response may be considerably degraded and trouble with stability of the amplifier may also be experienced. Since the frequency response required for measurements in a shear layer of given thickness increases directly proportional to the flow speed, it is difficult enough to provide an adequate frequency response for supersonic flow even at high overheat ratios. Therefore, at least up to the present, most measurements in supersonic flow have been made with constant-current apparatus.

It is fairly easy to make one's own constant-current anemometer, the most difficult part of the construction being the radio-frequency generator needed for setting the time constant. The standard way of doing this is to apply square-wave-modulated radio-frequency signals to the probe so that they appear as bursts of ohmic heating in the wire but do not pass through the audio-frequency amplifier: the time constant of the compensating amplifier is adjusted until the output most closely resembles a square wave (Fig. 29). Development of a constant-temperature ane-

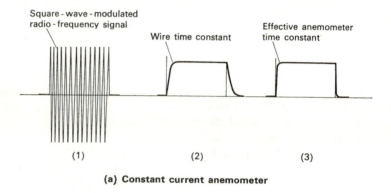

(a) Constant current anemometer

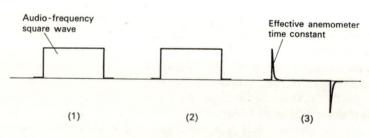

(b) Constant temperature anemometer

FIG. 29. Measuring the time constant of a hot-wire anemometer. (1) Signal applied to wire. (2) Signal entering amplifier. (3) Amplifier output.

mometer is more difficult and requires a good grasp of control theory: an audio-frequency square-wave generator is needed for checking the frequency response. The ratio of constant-temperature to constant-current sets on the commercial market probably reflects the relative ease of home construction rather than the relative popularity of the two systems although there is no doubt that the constant-temperature system has largely superseded the constant-current system for general laboratory and industrial use.

Occasionally, probes are operated in other modes (e.g. constant power dissipation) but this is done only for special reasons, connected with the electronics rather than the sensing element, and in the following discussion we shall consider only constant-current and constant-temperature operation.

The advantage of a hot *wire* over other heated elements is that it is self-supporting, so that heat conduction into the end supports does not greatly affect its behaviour and the frequency response is governed only by the thermal capacity of the wire, leading to a first-order linear differential equation and a simple time constant. However, a thin film, usually of platinum or nickel on a wedge-shaped or cone-shaped Pyrex support, is more robust than a hot wire and less liable to contamination by dirt, which usually adheres only near the front stagnation point: films are often preferred for measurements in liquids for these reasons and because they can be electrically insulated from the liquid by a layer of quartz. The basic one-dimensional analysis for the frequency response of an idealized film on a semi-infinite support ("substrate") is well documented, and the performance in the constant-temperature mode is quite adequate even though most of the heat is conducted into the substrate rather than into the stream. A serious difficulty that occurs in practice is that an appreciable fraction of the heat that enters the substrate is then conducted into the stream near the edges of the film rather than "disappearing" into the substrate: the fraction decreases to zero at high frequencies because the high-frequency temperature fluctuations do not penetrate far into the substrate, and the result is that, even when the film is operated at constant temperature, the sensitivity decreases from its steady-flow value, and asymptotes to a fraction thereof depending on the geometry of the probe and on the ratio of the thermal conductivity of the fluid to that of the substrate (see Section 5.1.1) but typically about 0·5 in air. At very high frequencies the response

decreases again according to the idealized theory. This effect practically
prohibits the use of films for quantitative work in gases unless some means
of dynamic calibration is available: liquids have higher thermal conductivity and the change in sensitivity with frequency is small enough to
be approximated by a theoretical expression or a once-for-all calibration
of a probe of similar design. A hot film on a thin cylindrical substrate
(a quartz fibre typically 0·1 mm in diameter) behaves in nearly all
respects like a hot wire: its advantage over a wire is that it can be quartz-
coated more easily. This type of probe seems to be the most satisfactory
for use in liquids, at least in principle, although the frequency response
may be affected by the finite time taken for *radial* conduction of heat
(effectively instantaneous in a thin metal wire): a rough calculation for
a 0·1-mm diameter fibre suggests that the response should be unaffected
below about 200 Hz, ample for most measurements in liquids. Most of
the remarks about films in the following chapters do not apply to this
type of probe: most of the remarks about hot wires do.

4.3. Doppler-shift Anemometers (Laser Anemometer, Sonic Anemometer)

If a train of waves of speed c and circular frequency ω is scattered by
an object moving at velocity U the frequency of the scattered waves
reaching a fixed observer is altered by the "Doppler shift"

$$\omega_d = \omega \frac{|U|}{c}(\cos(\phi_1 - \phi_2) - \cos\phi_1) = 2\omega \frac{|U|}{c}\sin\left(\phi_1 - \frac{\phi_2}{2}\right)\sin\phi_2$$

where ϕ_1 is the (variable) angle between the incident beam and the
velocity vector and ϕ_2 the (constant) angle between the incident beam
and the direction of the scattered waves reaching the observer (Fig. 30):
waves are usually scattered in all directions, although not uniformly,
so that ϕ_2 can be chosen independently). The "measured velocity
component", $|U|\sin(\phi_1 - \phi_2/2)$, bisects the obtuse angle between the
incident and scattered beams.

For this to be a practicable means of measuring velocity, whether
of solid bodies or of particles in a fluid flow, either the bandwidth $\Delta\omega$
of the incident radiation must be small compared to the frequency shift
to be measured or an interferometer technique must be used. If speeds
of the order of tens of metres per second are to be measured, the frac-

tional bandwidth $\Delta\omega/\omega$ must be small compared to $10/c$ where c is also in metres per second ($c \sim 3 \times 10^8$ for light). Ordinary light sources have a much larger bandwidth than this.

Sonic anemometers based on the Doppler shift of sound transmitted directly from a loudspeaker to a microphone ($\omega_d/\omega = U/c$ where U is the component of fluid velocity along the path of the sound waves) have been used for measurements in the atmosphere, particularly in the U.S.S.R. It is rather difficult to make the spatial resolution good enough for laboratory work. Sound wavelengths are much too large for the scattering technique to be used effectively.

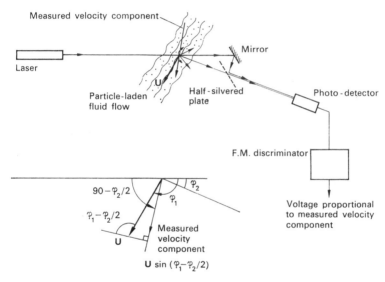

Fig. 30. Simplest optical arrangement of laser Doppler anemometer.

The output of typical continuous-wave lasers with a wavelength of roughly 6×10^{-7} m consists of several very narrow (10 Hz) spectral lines 10^8 to 10^9 Hz apart, giving an apparent fractional bandwith $\Delta\omega/\omega \approx 6 \times 10^{-6}$ (coherence length of the order of 10 cm). However, since the Doppler frequency shifts are *small* compared to $\Delta\omega$, the different lines can be treated as separate, highly coherent sources whose Doppler shifts differ by the (negligibly small) fraction $\Delta\omega/\omega$: therefore, providing that the Doppler-shift frequency can be measured directly, no trouble arises. The technique for obtaining an output at the difference frequency ω_d is to direct the incident light beam and the scattered light beam on to a square-law photo-detector: the output contains a component proportional to the square of the sum of the two wave amplitudes and this contains a low frequency ω_d, frequencies of the order of $\Delta\omega$ and of ω itself being smoothed out by the detector. Providing that the difference between the path lengths of the two beams is small compared with the coherence length, ω_d is not affected by the finite bandwidth: this simply means that the path lengths must not differ by more than 1 cm or so. A more stringent requirement is that the two beams must be nearly parallel when they reach the photo-detector, so that the relative phase shift across the width of the beams shall be small: this requirement is eased by focusing the two beams on the photo-detector surface so that their width is very small.

The photo-detector output at frequency ω_d (of the order of 1 MHz for $U = 1$ metre per second) is fed to a frequency-modulation discriminator which produces a voltage output proportional to the frequency: the simplest circuit that does this is one that emits a short pulse, of constant duration and amplitude, once per cycle of the input signal; when smoothed with a filter which will pass frequencies up to—say—$\omega_d/10$ only, the discriminator output voltage is proportional to the frequency ω_d, and thus proportional to U. An alternative is a "frequency tracker" which uses a phase-sensitive feedback loop to match the frequency of a voltage-controlled oscillator to the Doppler-shift frequency: the output is the control voltage itself.

To measure fluid flow velocity, it is usually necessary to add smoke to gases or dye to liquids in order to get enough scattered light: molecular scattering would not be suitable, even if the intensity were sufficient, because the frequency shift would depend on the total molecular velocity whose random part is far greater than the directed part due to the fluid flow.

Brownian motion of the smoke or dye particles may be appreciable, setting a lower limit to the acceptable particle size. "Modulation noise" appears in the discriminator voltage because the signal from a given scattering particle is received only while the particle is passing through the light beam and therefore consists of a finite train of waves. If there are more than a few illuminated particles in the field of view of the photo-detector the broadening of the Doppler frequency caused by the finite length of the wave trains is *independent* of the number of particles, because a statistical average is formed. The broadening, $\Delta\omega_d$ say, is of the order of the reciprocal of the particle passage time d/U where d is the beam width in the direction of **U**. Since $\omega_d \sim \omega U/c \sim U/\lambda$ we have $\Delta\omega_d/\omega_d \sim \lambda/d$ where λ is the wavelength of the *incident* light: if $\lambda = 6 \times 10^{-7}$ m and we require $\lambda/d < 0.01$, then $d > 60\,\mu$m (typical hot-wire diameters and lengths are $5\,\mu$m and $1000\,\mu$m respectively). The effects become more serious when **U** is nearly normal to the beam direction and ω_d is therefore small. The only way of reducing this frequency broadening, and consequent output noise, is to increase the number of cycles in each wave train by increasing the scattering volume or reducing the wavelength. In practice there is little control of the wavelength and one wishes to minimize the scattering volume so that the resolution will compete with that of the hot-wire anemometer. Of course a mechanical disadvantage of a small scattering volume is that setting up the optical system is more difficult. An idea of the problems is given by the fact that some investigators have chosen to traverse the measurement point across the flow by moving the test rig rather than the optical system: however, self-contained units that can be traversed as a whole are now commercially available. A further cause of broadening is the finite aperture of the beam, which leads to a change in the angle ϕ_1 as a particle passes through the beam. This can be minimized by suitable optical arrangements and by using a laser with a narrow output beam.

This recital of difficulties is intended to stop people throwing away their hot-wire anemometers and buying lasers: the advantages of having no solid probe in the flow are not usually enough to outweigh the expense and difficulty of the laser technique. However, many investigators are working on the laser anemometer for use in flows where hot wires would not survive, such as rocket nozzles: their work should lead to solution, or at least understanding, of the current problems, making the laser anemometer more attractive for general use. Therefore this account, updated in late 1974, may soon be out of date: the reader is referred to journals such as *Journal of Scientific Instruments* (*J. Phys. E.*) and *Review of Scientific Instruments* for more recent papers. Work done under NASA contracts appears in the NASA CR (Contractor Reports) series. For a recent review, see ref. 18.

Other Optical Techniques

Light absorption by dye or smoke can be used to measure concentration: calibration is difficult unless the properties of the tracer are carefully controlled. This subject is discussed in Chapter 7.

Measurements of attenuation along the whole length of a beam will strictly give only an average density or tracer concentration, but if the beam length is finite the exit intensity will still fluctuate. Fisher and Krause[19] argued that the correlation between the exit intensities of two beams which crossed within the flow would be representative of the conventional correlation area scale near the intersection point. Unknown calibration factors appear but, "to a useful approximation", the ratio of the exit intensity correlation measured with the crossed beams separated by a distance r to that measured with the beams intersecting is equal to the conventional correlation coefficient $R(r)$. Fisher and Krause used an extremely simple system with mercury vapour lamps producing beams of 2 mm diameter. A mist of water droplets was used in a jet flow at about 70 m s^{-1}.

A very detailed discussion of optical methods of measuring turbulence is given in ref. 20.

4.4. Glow-discharge[21] or Corona-discharge[22] Anemometers

Glow discharges (a blue glow at the cathode) can be maintained in air at atmospheric pressure with potentials of not more than a few hundred volts, if the electrodes are sharp and the gap between them is of the order of 100 μm: the current is of the order of 10 mA. The voltage E required to maintain a given current depends on the air speed U because the discharge takes place along a path of ionized gas which continually moves downstream and is replaced by freshly ionized gas. $\partial E/\partial I$ at constant U is negative so that a constant current source is necessary. $\partial E/\partial U$ at constant I varies with U and the calibration changes with time as the electrodes change their characteristics. In practice the drift seems to be considerably worse than that of a hot-wire anemometer, and the electrodes are probably sensitive to dirt in the air so that the device has little to recommend it for most purposes. The most recent use known to me was in a gas/solid suspension where a hot wire would not have survived and where a Doppler anemometer would have scattered indiscriminately from the solid particles and the tracer particles: the glow-discharge anemometer recorded the *gas* velocity fairly satisfactorily although considerable drift problems were reported.

A corona discharge can be generated with a higher voltage (several kV), a larger gap (several mm) and a smaller current (several μA): drift appears to be less serious than with the lower (and safer) voltages but this may depend more on the experimental conditions than the techniques.

Electrical discharges are sensitive to pressure which may be a further difficulty in wind-tunnel use.

At low gas pressures (less than 10^{-3} atm) electron beams can be used for mean or fluctuating density measurement. The fluorescent light output from a small section of the beam is focused on a photomultiplier whose output voltage is a roughly linear function of the gas density at the measurement point. Beam diameters of the order of 1 mm are easily obtainable so that the spatial resolution is comparable with that of the hot-wire or laser anemometer. Density can be deduced from beam attenuation at pressures up to about 10^{-2} atm but resolution is necessarily poorer. At these pressures, Reynolds numbers high enough for turbulence to appear are obtained only in hypersonic flow, to which the use of electron beams for density fluctuation measurements is therefore restricted.

4.5. The Pulsed-wire Anemometer (Fig. 31)

A very general means of measuring fluid velocity is to time the passage of a marked particle between two points. This has been done using solid particles[23], heat[24], foreign fluid or solute contaminant (a technique widely used in large-scale water flows is to inject a quantity of salt at one point and measure the electrical conductivity some distance downstream: radioactive contaminants have also been used). The most convenient technique for unsteady flows seems to be the heat-pulse method, using thin wires, normal to the direction of the required velocity component, both to introduce the heat and to detect the pulse. Bradbury[25] and Tombach[26] have recently developed the technique independently, one for flows with large changes in direction, the other for gas mixtures: the respective advantages of the pulsed-wire method in these two cases are that by having two "receiver" wires, one either side of the "transmitting" wire and at right angles to it, the true com-

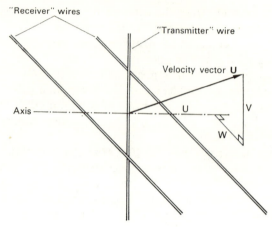

FIG. 31. Pulsed-wire anemometer (after Bradbury[26]). The measured velocity component is U.

ponent of velocity along the axis can be measured irrespective of sign; and that the velocity is measured directly and not by reference to any property of the fluid. Unfortunately the response is not linear (the actual measurement is of the time of flight, proportional to (velocity)$^{-1}$) so that measurements in highly turbulent flows require linearization: by recording the time of flight digitally for a large number of samples, the mean and mean square velocities can be worked out on a computer and linearized in the process. The frequency response, determined by the maximum permissible pulse repetition rate, is necessarily less than that of a conventional hot-wire anemometer, probably by an order of magnitude, but the spatial resolution need not be significantly worse. The pulse repetition rate may be limited by the spatial dimensions rather than the thermal inertia of the wire, because one pulse must be received before the next is transmitted to avoid confusion due to inductive pickup.

4.6. Particle Visualization

Photographs of a stream of dye or particles released from a fixed point or points give qualitative or even quantitative information about the velocity field, because the "Lagrangian" description of the motion of a marked particle is just as valid as the "Eulerian" description of the velocity at a point—although the Eulerian description is much easier to use in theoretical work and has therefore come to be preferred for experimental work also. A Eulerian description of the motion *can* be obtained by exposing the photograph for long enough for the particles to show up as short streaks, whose length and direction then give the velocity at different points. This technique is related to the "time-of-flight" methods discussed in Section 4.5.

A general technique for analysing pictures is to photograph them with a TV camera, whose "brightness voltage" output represents the illumination at each point on the successive lines in which the picture is scanned. Digital analysis of the tape-recorded brightness voltage can be used, in principle, to obtain any result obtainable by hand measurement of a photograph. To date, most of the analysis of flow visualization pictures has been done by hand, or, even more frequently, by eye. With experience, more useful information can be obtained by looking at the whole picture than by quantitative measurements at a few points (refs. 8 and 9 contain impressive results of this kind) but the interpretation of pictures of turbulence is often controversial. The best course for the beginner may be to abandon flow visualization in favour of quantitative point measurements as soon as controversy or uncertainty arise. This is not to deny the usefulness of visualization of the *mean* flow, as opposed to the eddies: standard techniques[27] can be used and the presence of turbulence brings difficulties of operation rather than interpretation.

4.7. Use of Steady-flow Techniques for Fluctuation Measurement

Nominally, Bernoulli's equation for an unsteady incompressible flow is

$$p + \tfrac{1}{2}\varrho\,|U|^2 + \varrho\,\frac{\partial\phi}{\partial t} = \text{const} = P, \quad \text{say,}$$

where U is the instantaneous velocity vector and the velocity potential ϕ is defined by $U = \text{grad }\phi$ or

$$U = \frac{\partial\phi}{\partial x}, \quad V = \frac{\partial\phi}{\partial y}, \quad W = \frac{\partial\phi}{\partial z}:$$

the equation is valid *only* for "potential" flow, with zero vorticity [if ϕ exists,

$$\frac{\partial V}{\partial x} - \frac{\partial U}{\partial y} = \frac{\partial}{\partial x}\left(\frac{\partial \phi}{\partial y}\right) - \frac{\partial}{\partial y}\left(\frac{\partial \phi}{\partial x}\right) = 0,$$

and similarly for the other components of vorticity]. What matters in practice is that the vorticity shall be small compared with $|U|/d$ where d is the distance, in the x direction say, over which we wish to apply Bernoulli's equation. Since in turbulent flow $\partial U/\partial y$, $\partial V/\partial x$ and other derivatives are all of the same order we can write this condition as $\partial U/\partial x \ll |U|/d$. Alternatively, differentiating Bernoulli's equation with respect to x, we see that the term in ϕ can be neglected if

$$\frac{\partial}{\partial x}\left(\frac{\partial \phi}{\partial t}\right) \ll \frac{1}{\varrho}\frac{\partial p}{\partial x} \sim \frac{1}{2}\frac{|U|^2}{d}.$$

But $\partial^2\phi/\partial x\,\partial t$ is $\partial U/\partial t$, so we require

$$\frac{1}{|U|}\frac{\partial U}{\partial t} \ll \frac{1}{2}\frac{|U|}{d}$$

which is of the same order as the first condition, since $\partial U/\partial t$ is of order $|U|\,\partial U/\partial x$. Therefore the restriction on the use of the steady-flow form of Bernoulli's equation for Pitot tube measurements in turbulent flow is the common-sense one that the flow over the instrument shall be quasi-steady, that is, that the velocity shall not change appreciably over a distance of the order of the tube diameter (since the length of the deceleration region ahead of the tube is not much larger than the tube diameter) or a time of the order of $d/|U|$. If this restriction is obeyed we can use the steady-flow calibration for yaw response as well.

The restriction on the use of static tubes is a little stronger: the velocity must not change appreciably over the distance between the front of the tube and the holes at which the pressure is measured. This implies that the conventional design of tube with the holes ten or more diameters from the front should be avoided: it is better to use a probe with the holes near the front, although its reading will be affected by the dynamic pressure. A further difficulty is that there is no type of static tube whose reading is independent of yaw angle in all directions over the range of roughly ± 15 deg required for measurements with $\sqrt{\overline{v^2}}/U$ up to 0·1. Therefore, even to measure static pressure alone, we need complete information about the magnitude and direction of **U**, obtained either from a Pitot-static yawmeter or from hot-wire measurements: the complete array of instruments must be small enough for the flow over it to be quasi-steady.

The diameter or other relevant dimension of the instrument must be small compared with the Kolmogorov length scale of the turbulence if the whole velocity or pressure-fluctuation spectrum is to be measured. This is impossible in the laboratory, and quite difficult even in the lower atmosphere: moreover, if the instrument is small its Reynolds number (based on the mean velocity) may be so small that the calibration depends strongly on Reynolds number. If the instrument diameter is equal to the Kolmogorov length scale $\eta \equiv (\nu^3/\varepsilon)^{\frac{1}{4}}$ then, in the logarithmic layer of a turbulent wall flow with $\varepsilon = u_\tau^3/Ky$ and, typically, $U = 20u_\tau$, we have

$$\frac{U\eta}{\nu} = 20\left(K\frac{u_\tau y}{\nu}\right)^{\frac{1}{4}}.$$

Suppose we require $U\eta/\nu = 200$: the required value of $u_\tau y/\nu$ is 25,000, obtainable with $y = 1$ metre and $u_\tau = 0.5$ m s^{-1}, corresponding to a strong wind close to the ground.

Even if we require only the *mean* total pressure and static pressure (to measure the mean velocity in a turbulent flow) we must still make the instrument small compared with the wavelengths containing most of the turbulent energy, *and* measure or guess the instantaneous yaw angle or its statistical properties. Since Pitot tubes are almost insensitive to yaw up to about ± 15 deg, we can assume with fair confidence that the mean pressure recorded by any fairly small pitot tube is the mean total pressure

$$p + \tfrac{1}{2}\varrho(|\mathbf{U}|^2 + \overline{u^2} + \overline{v^2} + \overline{w^2}).$$

The mean reading of a conventional static tube (holes 10 dia. from nose) varies strongly with size in the range of sizes likely to be used in laboratory shear flows: according to the results of ref. 28 in a highly turbulent jet, the reading is nearer to the actual mean static pressure than to the prediction of a quasi-steady analysis using the measured yaw calibration.

The conclusions are that mean total pressure can be measured with acceptable accuracy; that the fluctuating total pressure can probably be measured also, if the Pitot tube is small compared to the energy-containing eddies; but that neither the mean static pressure nor the fluctuating static pressure can be measured with any assurance of accuracy in laboratory flows. There has been some progress in making measurements of fluctuating pressure in the Earth's atmosphere: one of the difficulties is that of assessing the accuracy of the measurements.

In the atmosphere and ocean, the vane or windmill anemometers used for mean measurements are also suitable for fluctuation measurements if the flow is quasi-steady: in practice, instruments for fluctuation measurement may be made smaller than standard mean-flow instruments, at the expense of strength.

4.8. Measurement of Surface Pressure Fluctuations

The only fundamental difficulty in using a flush-mounted pressure transducer (microphone) of finite size is in calculating its spatial resolution. Assuming that the distribution of sensitivity over the face of the instrument is known, we still need to know rather more about the pressure-fluctuation field than we actually want to measure. The analogue of the Kolmogorov scaling used in Section 5.6 is, of course, inner-layer scaling, which does *not* imply isotropy of the surface pressure fluctuation field but which does enable us to write

$$\Phi_{\text{true}} = \frac{\varrho^2 u_\tau^4}{k_1} f\left(\frac{k_3}{k_1}, \frac{u_\tau}{k_1 \nu}\right)$$

and

$$\frac{\Phi_{\text{measured}}}{\Phi_{\text{true}}} = f\left(k_1 d, \frac{k_3}{k_1}, \frac{u_\tau d}{\nu}\right)$$

where $\Phi(k_1, k_3)$ is the two-dimensional spectral density, k_1 and k_3 the wave numbers in the x and z directions along the surface, and d the diameter of the transducer. For non-circular transducers geometrical parameters also appear. $u_\tau d/\nu$ may be important for wave numbers typical of the pressure fluctuations generated in the viscous sublayer ($u_\tau y/\nu < 30$, say) but as $u_\tau d/\nu$ is usually *much* greater than 30 the transducer will not respond to these fluctuations ($f \simeq 0$) and $u_\tau d/\nu$ may be omitted. Moreover, if only a one-dimensional spectrum is required we have

$$\frac{\phi_{\text{measured}}}{\phi_{\text{true}}} = f(k_1 d)$$

where the universal function can be obtained from once-for-all measurements with a series of geometrically similar transducers. The analysis is valid for the range of wave numbers typical of the pressure fluctuations generated in the inner layer ($y < 0.2\delta$ in a boundary layer, leading to $k_1\delta < 5$, say). Unfortunately, one of the main applications of surface pressure fluctuation measurements is to excitation of surface waves, so one also wishes to correct wave-number/frequency spectra for transducer resolution: the extra dimensionless group $\omega d/U_c$ appears in the correction functions above, where U_c is the convection velocity of the pressure pattern (some suitable generalization of the phase velocity: see Section 2.7). For a pure frequency spectrum

$$\frac{\phi_{\text{measured}}}{\phi_{\text{true}}} = f\left(\frac{\omega d}{U_c}\right)$$

if $u_\tau d/\nu$ is so large that $f \simeq 0$ for frequencies generated in the viscous sublayer. This is the form in which most experimental results have been presented.[29]

Capacitor microphones, commercially available in diameters down to $\frac{1}{4}$ in. (0.63 cm), piezoelectric crystals such as barium titanate and lead zirconate–titanate, and probe microphones connected by a tube to a hole in the surface have all been used. Crystals have been made as small as 0.07 cm diameter but the smallest nominal transducer diameters are obtained with probe microphones, commercial or home-made: hole diameters as small as 0.02 cm have been used. The distribution of sensitivity over the area of the hole is unknown and there may be some response, via the fluid flow, to pressures applied at a short distance from the hole: this makes corrections for the effect of transducer size conjectural. Another disadvantage of probe microphones is that the signal is delayed by the time taken for a sound wave to travel the length of the probe tube.

4.9. Specialized Techniques of Turbulence Measurement

This section is an annotated list of references to techniques in which either the principle, or the quantity to be measured, or the flows to which the technique is to be applied, is specialized. By virtue of this, these are not alternatives to the hot wire for turbulence measurements in general, but they have been shown to be effective in special cases.

Techniques allied to the hot wire

ELLINGTON, D. and TROTTIER, G. Some observations on the application of cooled-film anemometry to the study of the turbulent characteristics of hypersonic wakes, *Canadian Armament Research and Development Estab.*, *CARDE T.N.* 1773/67 (1967).

HARRIS, G. S. Further development of the cold tip velocity meter, *J. Sci. Instrum.* Ser. 2, **2**, 83 (1969).

The cooled-element anemometer is useful if the fluid temperature is too near boiling point or the maximum permissible heated-element temperature: commercial probes are available.

DAS, M. M. Extended application of a single hot-film probe for the measurement of turbulence in a flow without mean velocity, *Univ. of Calif., Hydraulic Engng. Lab. Rept.* HEL-2-20, AD-681923 (1968).

A relative velocity is provided by oscillating the probe: again, a commercial probe is available.

WILLS, J. A. B., A traversing orifice-hot-wire probe pair for use in wall pressure correlation measurements, *J. Sci. Instrum.*, Series 2, **1**, 447, (1968).

This is a development of an idea due to Kovasznay: the hot wire is used as the sensing element in a small-diameter probe microphone.

Velocity from pressure measurements

JEZDINSKY, V. Measurement of turbulence by pressure probes, *A.I.A.A. Journal* **4**, 2072 (1966).

A twin-pitot yawmeter is used for the measurement of lateral component fluctuations.

KAMOTANI, Y. and WISKIND, H. K. Measurement of turbulence quantities by pressure probes, *N.A.S.A. Contractor Rept.* no. CR-99336 (1968).

This is a comment on, and an extension of, Jezdinsky's method of data reduction.

SIDDON, T. E. and RIBNER, H. S. An aerofoil probe for measuring the transverse component of turbulence, *A.I.A.A. Journal* **3**, 747 (1965).

The instantaneous lift of a very small aerofoil (a function of the incidence) is measured by strain gauges.

TRITTON, D. J. The use of a fibre anemometer in turbulent flows, *J. Fluid Mech.* **16**, 269 (1963).

SLEATH, J. F. A. A device for velocity measurements in oscillatory boundary layers in water, *J. Sci. Instrum.*, Ser. 2, **2**, 446 (1969).

The drag of a cylinder is a function of velocity: Tritton, concerned only with very-low-frequency flows, used a microscope to observe the bending of a cantilevered fibre; Sleath used strain gauges on the fibre supports.

Observation of particles

DAVIS, W. and FOX, R. W. An evaluation of the hydrogen bubble technique for the quantitative determination of fluid velocities within clear tubes, *J. Basic Engng.* **89D**, 771 (1967).

The application is to mean velocities but the principles apply to turbulence measurement also.

CLARKE, W. B. Liquid velocity measurement by electronic detection of particle images, *M.I.T. Dept. of Chem. Engng. Progress Rept.*, Contract No. 3963 (10) (1968).

JONES, B. G., CHAO, B. T. and SHIRAZI, M. A. An experimental study of the motion of small particles in a turbulent field, using digital techniques for statistical data processing, *Developments in Mechanics* **4** (10th Midwestern Mechanics Conference, Colorado State Univ., 1967), p. 1249.

Droplet-laden gas flows

GOLDSCHMIDT, V. W. and ESKINAZI, S. Two-phase turbulent flow in a plane jet, *J. Appl. Mech.* **33E**, 735 (1966).

Droplet impingement on the hot wire produces a high-frequency spike which can be distinguished from the turbulence signal.

Miscellaneous measurement principles

MITCHELL, J. E. and HANRATTY, T. J. A study of turbulence at a wall using an electrochemical wall shear stress meter, *J. Fluid Mech.* **26**, 199 (1966).

Some very useful information about the viscous sublayer has been obtained with this "mass transfer" technique: the advantage over hot films imbedded in the surface is that mass transfer takes place only from very near the electrode itself, whereas heat transfer through the substrate increases the effective size, and decreases the effective aspect ratio, of the hot film.

MAGRINI, U. and PISCONI, C. On the evaluation of "Liquid Fluctuating Velocity" during bubbling of gas through a liquid, *International J. Heat and Mass Transf.* **12**, 1325, 1969 (see also *Rev. Sci. Instrum.* **37**, 627 (1966)).

This is an electrolytic flowmeter for very low speeds (less than 15 cm s^{-1}).

McCORMICK, W. S. and BIRKEMEIER, W. P. Optimum detectors for the NMR flowmeter, *Rev. Sci. Instrum.* **40**, 346 (1969).

"Time of flight" method akin to the pulsed-wire anemometer but using nuclear magnetic resonance to trace the fluid.

WALLACE, J. E. Hypersonic turbulent boundary layer measurements using an electron beam, *A.I.A.A. Journal* **7**, 757 (1969).

Recent flow-visualization techniques

MILLER, E. B. The visualization of turbulent flows by means of flow birefringence *N.A.S.A. Contractor Rept.* no. CR-88703, 1967.

This is one of the latest references to the technique: a difficulty is that the suspensions that exhibit birefringence also seem to exhibit non-Newtonian behaviour.

A general review of flow-visualization techniques, including some intended primarily for turbulent flow, is given in ref. 27. Techniques for quantitative measurement of smoke, dye or contaminant concentration are discussed in Chapter 7.

CHAPTER 5

The Hot-wire Anemometer

("That's exactly the method", the Bellman bold
In a hasty parenthesis cried,
"That's exactly the way I have always been told
That the capture of Snarks should be tried!")

WIRES for turbulence measurement are usually made of platinum or tungsten with diameters of the order of 5 microns (Fig. 32). The examples given in Table 1 are all for a 5-micron tungsten wire having a length of 1 mm, an electrical resistance $R_a = 3{\cdot}5$ ohms at ambient temperature and $R_w = 7$ ohms when heated, and with its axis in the z direction unless otherwise stated. We shall consider a typical air flow, with a speed of 30 m s^{-1} in the x direction and at normal temperature and pressure, giving a wire Reynolds number based on diameter of about 10, a heat-transfer Nusselt number† of about 2, and a time constant of about 0·6 millisecond. The wire dimensions and material are not, of course, critical. Pure metals are preferred because they have higher temperature coefficients of resistance than alloys and tungsten is becoming the most popular material, although it is difficult to handle, because it is much stronger than any other convenient material. Platinum is also widely used, principally because it can be obtained in the convenient form of Wollaston wire. In the Wollaston process the thin platinum rod to be drawn is covered by a thick sheath of silver (or some other metal with about the same Young's modulus and yield stress) and the assembly is drawn like a solid billet down to the smallest practicable outside diameter so that the diameter of the platinum core is a small fraction

† $\dfrac{\text{heat loss/unit length}}{\pi k_f (T_w - T_f)}$: suffix w = wire, f = fluid

TABLE 1. PROBE MATERIALS (WHERE POSSIBLE, PROPERTIES FOR THIN WIRES HAVE BEEN QUOTED)

	Breaking stress N cm^{-2}	Maximum temperature	Soft-solderable?	Weldable?	Available as Wollaston wire?	Minimum diameter available	Temperature coefficient of resistivity α per deg. C	Resistivity ohm–cm at 0°C	Thermal conductivity k, W/cm deg. C at 0°C
Tungsten	250,000	300°C (oxidizes)	No	Yes if plated	No	2·5 μm	0·0040	4·9 × 10^{-6}	1·9
Platinum	30,000	1200°C (melting point 1750°C)	Yes	Yes	Yes	1 μm	0·0038	9·8 × 10^{-6}	0·7
Platinum–Rhodium (90/10)	60,000	1400°C (melting point 1600°C)	Yes	Yes	Yes	0·000025 in. (0·6 μm)	0·0016	19 × 10^{-6}	0·4

Platinum–iridium alloys are unstable at high temperatures and should generally be avoided.

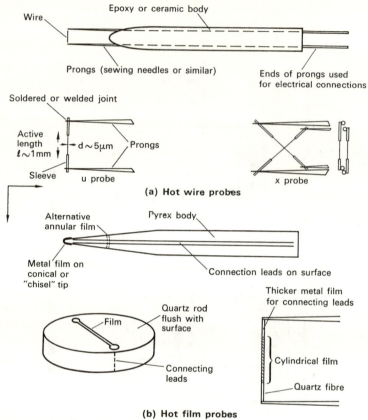

(a) Hot wire probes

(b) Hot film probes

FIG. 32. Probe designs.

of that otherwise obtainable: an additional advantage in use is that most of the manipulation can be done before etching the silver sheath from the wire and exposing the small and fragile platinum core. Platinum–rhodium alloy, which is stronger than platinum but has a smaller temperature coefficient of resistance, is also available in Wollaston wire, but there is no suitable material for sheathing tungsten; the smallest diameters are obtained by etching. Platinum-plated tungsten is less liable to oxidation than pure tungsten and is also easier to weld or solder.

Films are usually platinum or nickel, although most metals can be sputtered. Pyrex or similar glass is the universal material for the support (substrate).

5.1. Heat Transfer

The electrical heat input to the wire is transferred out of the wire by radiation, buoyant convection, conduction along the wire to its end supports and forced convection by the fluid flow. The radiative heat loss from our typical wire is only about 0·1 per cent of the electrical input and can be neglected, but it may be important in very low density flows. Buoyant convection is important only at very low speeds and a criterion for its neglect, adapted from Collis and Williams[30], is that the Reynolds number should be greater than twice the cube root of the Grashof number,† which is the ratio of a typical buoyancy force to a typical viscous force. For our typical wire, the Grashof number is about 6×10^{-6} so that buoyant convection can be neglected for fluid speeds greater than 5 cm/sec. Heat loss by buoyant convection depends on the inclination of the wire to the vertical but this dependence is fairly slight because the actual heat transfer from the wire surface takes place entirely by conduction and, in the case of fine wires, buoyant convection becomes important only at many diameters from the wire where the effective Grashof number becomes of order unity, so that the apparent aspect ratio of the body from which buoyant convection takes place is very much less than that of the wire itself. For most purposes, we can neglect radiation, buoyant convection and also the various small thermoelectric effects that occur: the main contributions to heat transfer from the wire are conduction to the supports and forced convection to the fluid flow, which we proceed to discuss because, even if probes are to be individually calibrated, some knowledge of the physics of heat transfer is necessary to assess the reliability of the calibration.

5.1.1. Conduction to the Supports

The supports are chosen to be much thicker than the wire, partly for reasons of strength and partly so that they shall not be heated appreciably by the electric current. Therefore the temperature at the ends of

† $\dfrac{gd^3(T_w - T_f)}{v^2 T_f}$.

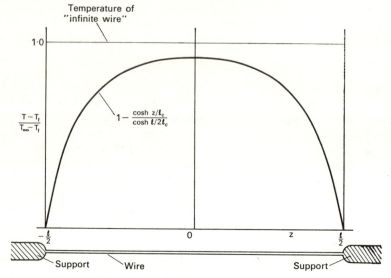

FIG. 33. Effect of supports on wire temperature distribution.

the wire is very nearly the temperature of the fluid, T_f, and conductive heat transfer along the wire to the supports is appreciable. If we assume for simplicity that the convective heat transfer to the fluid is directly proportional to the difference between T_f and the local wire temperature T, we get a second-order differential equation for T in terms of the distance z from the mid-point of the wire. The solution[31] is, with various approximations,

$$\frac{T - T_f}{T_\infty - T_f} = 1 - \frac{\cosh z/l_c}{\cosh l/2l_c}$$

where T_∞ is the temperature an infinitely long wire would reach and l_c, called the "cold length" because the total resistance is that of a wire of length $l - 2l_c$ and temperature T_∞, is defined by

$$l_c = \frac{d}{2} \sqrt{\left/ \left(\frac{R_w}{R_a} \frac{k_w}{k_f} \frac{1}{\text{Nu}} \right) \right.},$$

Nu being the forced-convection Nusselt number (Section 5.1.2). The solution becomes accurate as $R_w/R_a \rightarrow 1$: l_c is overestimated by taking the measured (average) R_w/R_a but this is conservative. For our typical tungsten wire with $R_w/R_a = 2$, l_c is about $30d$. If l/d is 200, the average temperature difference $T_w - T_f$ is about $0·7 (T_\infty - T_f)$ and the temperature difference in the middle of the wire is $0·93 (T_\infty - T_f)$ or $1·3$ times the average temperature difference (Fig. 33). The ratio of conductive heat transfer to both supports, H_s, to convective heat transfer to the fluid, H_f, is roughly

$$\sqrt{\left(\frac{R_a}{R_w}\frac{k_w}{k_f}\right)}\frac{d}{l}\frac{1}{\sqrt{Nu}}$$

or about $0·15$ for our typical tungsten wire: tungsten has a much higher thermal conductivity than platinum and its alloys. The difficulties caused by the non-uniform temperature distribution are:

(1) The temperature in the middle of the wire is higher than the temperature inferred from the wire resistance, so that oxidation of the wire material in air, or bubble formation in liquids, occurs sooner than expected. This difficulty is trivial once one appreciates that there *is* a difficulty.

(2) Calibration constants for heat loss by forced convection depend on the length of the wire: this is one of many reasons for distrusting universal calibrations.

(3) The response of the wire temperature T_w to fluctuations in heat transfer involves a redistribution in temperature because l_c changes. A typical time for this redistribution to take place is obtained from dimensional analysis as $l_c^2\varrho_wc_w/k_w$, where c_w is the specific heat of the wire material, and for tungsten wire with $l_c = 30d = 0·015$ cm this is about $0·3$ msec, roughly the same as the response time of an infinite wire due to its thermal capacity. However, the redistribution of temperature has very little effect on heat transfer by forced convection because the latter is, to a first approximation, proportional to the temperature difference, so that the redistribution of temperature leaves the total convective heat transfer for a given average temperature unaltered and the main effect of the finite response time is on the conductive heat transfer to the supports.

Because of the inverse square root dependence of H_s/H_f on Nusselt number, the ratio of the change in conductive heat transfer to the change in convective heat transfer produced by a given change in mean fluid speed is $-\frac{1}{2}(H_s/H_f)$, say -0.075, and the maximum change in frequency response, occurring at frequencies many times the reciprocal of the response time, is less than this because the slowness of response of the conductive heat transfer leads to transient changes in local wire temperature which produce compensating changes in convective heat transfer. Therefore, the effect is not serious in practice except for very short wires (and even there it is likely to be no larger than the effect of aerodynamic interference from the supports which also depends on frequency or, strictly, eddy size). In the case of non-cylindrical hot films, where the conductive heat transfer to the support is nominally uniformly distributed over the area of the film, the compensating effect mentioned above does not occur.

It is difficult to make useful theoretical corrections to the frequency response and phase lag because the transient temperature distribution is governed by a complicated differential equation, and the usual practice is to use wires that are long enough ($l/d > 200$) for the effects on the dynamic response to be small. In the case of hot films, the effects are much larger, but the theory is much simpler, providing that a few approximations are made: Bellhouse and Schultz[32] and Bellhouse and Rasmussen[33] have shown that the frequency response is a function of a modified Nusselt number and a dimensionless frequency,

$$\frac{k_f}{k_s}\,\mathrm{Nu} \equiv \frac{lH_f}{Ak_s(T_w - T_f)} \quad \text{and} \quad \frac{\omega l^2}{k_s}$$

where l is a typical thickness of the *support*, A is the area of the film and k_s is the thermal conductivity of the *support* material. At large ω the response becomes independent of frequency but still depends on $(k_f/k_s)\mathrm{Nu}$. Typical response measurements are given in ref. 32. At even higher frequencies the response of a film decreases once more because of conventional thermal-capacity effects, calculable by a one-dimensional analysis: the response does *not* have a simple "time-constant" character.

5.1.2. *Forced Convection*

Heat transfer from hot wires is usually expressed as a dimensionless Nusselt number, equal to the rate of heat transfer to the fluid per unit area, H_f/A, divided by the product of the thermal conductivity of the fluid k_f and a typical temperature gradient. Taking the typical temperature gradient as $(T_w - T_f)/d$ the Nusselt number becomes

$$Nu \equiv \frac{H_f/l}{\pi k_f(T_w - T_f)}.$$

The diameter vanishes because it appears both in the surface area and in the typical temperature gradient chosen: in the case of a film or other non-cylindrical element

$$Nu = \frac{lH_f}{Ak_f(T_w - T_f)}.$$

The parameter normally used in discussions of fluid-dynamic heat transfer is the Stanton number (Section 3.10):

$$St = \frac{H_f/\varrho c_p \pi l d}{U(T_w - T_f)} = \frac{H_f/l}{\pi k(T_w - T_f)} \bigg/ \left(\frac{\varrho Ud}{\mu} \frac{\mu c_p}{k} \right) \equiv Nu/(Re . Pr).$$

At low speeds and moderate densities the Nusselt number (or Stanton number) depends on the Reynolds number, the Prandtl number and the angle of inclination of the fluid stream to the wire. In what follows we shall base the Reynolds number on the resultant velocity in a plane perpendicular to the wire axis and denote the angle of inclination of the velocity vector to this plane by ψ.

The Prandtl number, unlike the other dimensionless numbers used here, is a property of the fluid. It is very nearly constant in gases, being about 0·7 for air (unity according to the simple kinetic theory of gases) and need not be considered explicitly in the calibration. In liquids, it varies considerably with temperature because μ depends upon temperature whereas k and c_p do not. For instance, in water the Prandtl number decreases from 7 at 15°C to 1·7 at boiling point, the decrease at low temperatures being about 1 per cent per degree centigrade. Little systematic work has been done on the effect of Prandtl number on hot-wire calibration although several conflicting empirical correlations exist. Corrsin[31] quotes the results of Kramers and van der Hegge Zijnen who found fair agreement with

$$Nu = 0·42 \, Pr^{0·20} + 0·57 \, Pr^{0·33} . Re^{0·50}$$

for $0.71 < \text{Pr} < 1000$ and $0.01 < \text{Re} < 1000$ (for air, $\text{Pr} = 0.71$: this formula is a rough approximation to Collis and Williams' results in the range $\text{Re} < 140$). Bourke *et al.*[34] adopted

$$\text{Nu} = 0.75 \, \text{Pr}^{0.20} + 0.67 \, \text{Pr}^{0.9} \cdot \text{Re}^{0.50}$$

for $0.85 < \text{Pr} < 1.6$ and $\text{Re} < 20$ in carbon dioxide near the critical point but this correlation is not necessarily applicable in general and appears to include heat conduction to the supports as well as forced convection (the hot wire was similar to the "typical" wire). Several authors assume $\text{Nu} = f(\text{Re}.\text{Pr})$, where $\text{Re}.\text{Pr} = \varrho c_p U d/k_f$ is the Péclet number, but there is no satisfactory justification for this assumption, which arose historically from King's assumption that the velocity field was independent of Reynolds number. Individual calibration seems advisable at present.

In gases, the calibration is independent of the Knudsen number, $\text{Kn} = (\text{mean free path})/d$, if $\text{Kn} < 0.015$ ($d > 4 \, \mu\text{m}$ at atmospheric pressure). Since Kn depends only on pressure it need not be considered explicitly in low-speed flows where the absolute pressure is nearly constant.

In non-Newtonian fluids, the flow round fine wires and other small probes may be even more seriously affected than the large-scale flow: in polymer solutions, the concentration must be added to the list of variables above.

We finally come to the variation of Nusselt number with Reynolds number. This is, of course, the main information required for hot-wire anemometry and has been explored by many workers, among the first being King (see ref. 31) who gave the semi-empirical form, $\text{Nu} = A + B \, \text{Re}^{0.5}$ in the present notation, known as King's law. More recent work has shown that an exponent of 0.45 gives a better correlation than an exponent of 0.5 in the Reynolds number range usual in hot-wire ane-mometry. Like power-law skin-friction formulae, these correlations are applicable only over a limited range of Reynolds number and there is no physical reason why a single exponent should do for an indefinitely large range. Collis and Williams[30] chose an exponent of 0.45 to give the best fit in the range $0.02 < \text{Re} < 44$: they used very long wires and the higher Reynolds numbers were obtained with thicker wires, still at a very low Mach number. Calibrations of short, fine wires like our typical anemometer wire show a slight variation of exponent with speed: it is not clear whether this would appear for long wires at low Mach number if the Collis–Williams experiments were repeated with even greater care, or whether it is attributable to end losses, prong interference and Mach number effects. Large variations of exponent with speed occur if the intercept A in the formula $\text{Nu} = A + B\,\text{Re}^n$ is obtained from the buoyant heat transfer at zero speed rather than from a best fit to the forced-convection measurements.

The temperature difference between the wire and the fluid is usually large enough for the choice of reference temperature for the fluid properties to be in doubt. Collis and Williams evaluated fluid properties at $T_m \equiv (T_w + T_f)/2$ but found that an additional "temperature loading" factor had to be applied to the Nusselt number. Their final result, probably the most accurate yet obtained, is

$$\text{Nu}\left(\frac{T_m}{T_f}\right)^{-0.17} = A + B\,\text{Re}^n$$

	$0.02 < \text{Re} < 44$	$44 < \text{Re} < 140$
n	0·45	0·51
A	0·24	0
B	0·56	0·48

Davies and Fisher[35] present results that apparently disagree strongly with those of Collis and Williams: Bradbury and Castro (*J. Fluid Mech.*, **51**, 487, 1972) show that the two sets of results agree when Collis and Williams' reference temperature and temperature-loading factor are used for both. The search for a reference temperature that will collapse all results without a temperature loading factor is futile: the effective reference temperature depends on weighted integrals of the density and thermal conductivity through the shear layer round the wire. Not only would slightly different reference temperatures be needed for ϱ and for μ or k but they would, like other properties of the shear layer, depend on the Reynolds number and the Prandtl number, and the Reynolds number and Prandtl number would depend on the reference temperature. In practice, moreover, the wire temperature varies considerably along its length. In constant temperature operation, the effect of temperature loading does not appear unless the fluid temperature changes.

5.1.3. Calibration of Hot Wires

There is a divergence of opinion among users of hot wires about the usefulness of universal correlations for heat transfer like those discussed in the last section (users of hot films have no such correlations available at present, although there is nothing in principle to prevent correlations being established for a few standard probe shapes). My own opinion, firmly based on experience, is that universal calibrations like King's law or the 0·45 law give a useful guide to the best way of plotting results, but that uncertainties about the properties of the probe and the fluid are such that individual calibration of probes is desirable if good accuracy is to be obtained. Of course, the thoroughness with which individual calibration is done can be chosen according to the consistency and quality of probe manufacture and the variability of fluid properties: for instance, one would rarely trouble to find the optimum value of n for each probe but would use a fixed value chosen from experience with probes of the same type in similar conditions. Nominally a King's law line, with chosen n, can be defined by two calibration points: one frequently finds it suggested that one point can be taken at zero speed, but the buoyant-convection heat loss at zero speed depends on the Grashof number, which has no direct connection with the Reynolds number. It happens that for wires of the size normally used in turbulence measurements in air there is not much difference between the actual heat loss at zero speed and an extrapolation of the forced-convection law, but the ratio of the two varies with fluid temperature and my own experience is that, where accurate answers are required, it is unwise to use the still-air heat transfer as a point on the forced-convection calibration curve. Of course, the error involved decreases as the speed at which turbulence measurements are to be made increases.

It is possible to ignore the universal calibration laws altogether and simply plot wire voltage E against speed U (measured, say, with a Pitot tube and static tube near—but not too near—the wire). The graph of E against U can then be differentiated, graphically or numerically, to give dE/dU, which is the quantity actually needed to convert values of voltage fluctuation into values of velocity fluctuation. However, if one is going to fit a curve to the points to obtain the slope, one might as well fit a King's law or 0·45 law—that is, plot $I^2 R_w/(R_w - R_f)$, which is a fixed factor times the Nusselt number, against $U^{0·45}$ or $U^{0·5}$, U being a fixed factor times the Reynolds number. The result should of course be a straight line. For the most usual cases of constant current or constant temperature operation, the "Nusselt number" can be further simplified to $R_w/(R_w - R_f)$ or E^2 respectively. An objection to this procedure is that if the calibration does not follow a 0·5 or 0·45 power law one is tempted to force it to do so rather than drawing the best curve through the points: the answer is that a probe whose calibration differs appreciably from that of similar probes calibrated previously should be discarded; if one's probe calibrations consistently follow, say, a 0·55 power law, then this index should be used in plotting instead of 0·5 or 0·45— assuming that one is happy about the probe design and the calibration technique.

Once a given wire has been calibrated, repeats of only two or three points should suffice for a check because it is found in practice that the calibration line usually remains straight throughout the life of the probe, merely altering its slope and intercept for reasons discussed below (Sections 5.2, 5.4).

5.2. The Effect of Fluid Temperature

One of the main difficulties in the use of hot wires is the effect of fluid temperature changes on the heat transfer. In air, 1 °C temperature change may produce an error of 2 per cent in measured velocity. Not only does the temperature difference between the wire and the fluid appear explicitly in the Nusselt number but the thermal conductivity and the viscosity of the fluid also appear in the heat transfer relation, and these are both functions of temperature. In air, the thermal conductivity and the viscosity both vary as $T^{0·76}$ approximately, and the density is inversely proportional to temperature: this means that the variation of Nu and the variation of $Re^{0·45}$ are very nearly the same, assuming that the same reference temperature is used for both, so that the hot-wire calibration can be written as

$$\frac{H_f}{T_w - T_f} \propto \frac{I^2 R_w}{R_w - R_f} = A_1(T_f) + B_1 U^{0·45}$$

to a fair approximation, for a given wire, where A_1 is a function of T_f but B_1 is not. Therefore, providing that the ratio of the wire temperature to the fluid temperature, T_w/T_f, is maintained constant (i.e. R_w/R_a = constant) the calibration shown in Fig. 34 will alter much less in slope than in intercept as the air temperature changes. It is therefore fairly easy to allow for small variations in air temperature, particularly in the case of turbulence measurements, where, for example, the ratio of r.m.s. turbulence intensity to local mean velocity can be obtained from the above equation as

$$\frac{\sqrt{\overline{u^2}}}{U} = \frac{\sqrt{\overline{h^2}}}{0.45 B_1 U^{0.45}}, \quad \text{where } h \text{ is the fluctuating part of } \frac{I^2 R_w}{R_w - R_a}.$$

This can be evaluated irrespective of changes in temperature if U is known and if we assume that the slope B_1 of the heat transfer curve is unaltered. In other fluids, where the variation of fluid properties with temperature is less convenient or in cases where large temperature

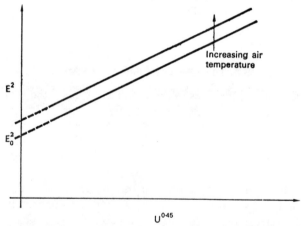

FIG. 34. Effect of air temperature on wire calibration: $R_w/R_a = $ constant.

changes are expected, it may be necessary to devise more complicated schemes: the general principle is to do a calibration at one fluid temperature and then to use our theoretical or empirical knowledge of the variation of heat transfer with fluid temperature to deduce calibrations at another temperature (e.g. ref. 36 and Chapter 7).

The usual cause of a change in a fluid temperature during a run is power dissipation in the test rig. The problem is particularly serious for small closed-circuit wind tunnels or water tunnels. It is sometimes possible to fit the test rig with a refrigerator but the only palliative normally available is to run the test rig, before taking the measurements, for a sufficient time for its equilibrium temperature to be reached. This process can be speeded up by running at maximum speed and then reducing speed to the value at which measurements are to be made. If the mean temperature of the fluid is changing rapidly with time or changing with distance across the test rig, then it is highly probable that temperature fluctuations will occur in the flow. Because the hot wire responds to these as well as to velocity fluctuations, turbulence measurements will be in error unless some allowance is made, and since the process of extracting velocity *and* temperature fluctuations from hot-wire measurements is rather complicated, it is worth going to some trouble to suppress temperature fluctuations within the fluid. No general rules can be given for doing this except to point out that if the test rig is fitted with a heat exchanger, it should *not* be situated just upstream of the working section. If temperature fluctuations are inevitable, as in the case of measurements in the atmosphere or in ventilation ducting, their effect on the velocity fluctuation measurements can be minimized by running the wire at as high a temperature as possible. This subject is discussed in more detail in Chapter 7.

The voltage/velocity calibration can be made independent of fluid temperature, at least over a small range, by inserting a temperature-sensitive element, exposed to the fluid, in another arm of the bridge of a constant-temperature anemometer or the heating circuit of a constant-current anemometer. A commercial probe, with a temperature-compensating thermistor element mounted just behind the wire, is now available: the compensating element has a long time constant and therefore does not compensate for short-period temperature fluctuations.

5.3. The Effect of Flow Direction

5.3.1. *Flow in the Plane Normal to the Wire Axis*

A cylindrical hot wire is entirely insensitive to changes of flow direction in the plane normal to the axis of the wire. Supposing for the moment that the instantaneous component of velocity along the axis of the wire,

$W + w$, has no effect we see that the wire will respond to $\sqrt{\{(U + u)^2 + v^2\}}$, not $U + u$. If the ratio of the fluctuating components to the mean velocity is small, this can be expanded as $U\{1 + (u/U) + \frac{1}{2}(v^2/U^2)\}$, of which the fluctuating part is $u + \frac{1}{2}\{(v^2 - \overline{v^2})/U\}$. The effect of these higher-order terms depends on the quantity being measured and, of course, upon the intensity of the turbulence. In practice, of course, the wire supports are asymmetrical in this plane and the wire calibration will depend slightly on the inclination of the probe body to the flow direction. Measurements for a typical commercial probe (ref. 37) gave an "effective velocity" $\sqrt{\{(U + u)^2 + 1\cdot3v^2\}}$ for large-scale velocity variations; the response to wavelengths smaller than the probe body length would differ less from $\sqrt{\{(U + u)^2 + v^2\}}$ because body interference would be negligible.

From above, the ratio of measured $\overline{u^2}$ to true $\overline{u^2}$ is

$$\left(1 + \frac{\overline{uv^2}}{U\overline{u^2}} + \frac{(\overline{v^4} - (\overline{v^2})^2)}{4U^2\overline{u^2}} + \cdots\right):$$

now $\overline{v^4}/(\overline{v^2})^2$ can be taken as 3 (it takes much higher values in the intermittent region of a shear flow but the intensity is small there) and $\overline{uv^2}/\sqrt{\overline{u^2}\,\overline{v^2}}$ is roughly the same as $\overline{u^3}/(\overline{u^2})^{3/2}$ (<0 near the edge of a boundary layer), so that the factor becomes of order

$$1 + \left(\frac{\overline{v^2}}{U\sqrt{\overline{u^2}}}\right)S + \frac{1}{2}\left(\frac{\overline{v^2}}{U\sqrt{\overline{u^2}}}\right)^2 + \cdots$$

where $S = \overline{u^3}/(\overline{u^2})^{3/2} \simeq -2(y/\delta)^2$ in a boundary layer. Taking $\overline{v^2} = \overline{u^2}$ (pessimistic) and $\sqrt{\overline{u^2}}/U = 0\cdot1(1 - y/\delta)$, the factor on $\overline{u^2}$ is at worst $0\cdot97$ (at $y/\delta \simeq 0\cdot6$) and the second term is always negligible. In jet flows, S is of the same order but $\sqrt{\overline{u^2}}/U$ may reach $0\cdot3$ or more, and the corrections become rather large and uncertain.

It is important to notice that we have said nothing of the non-linearity of the hot-wire calibration itself; this sort of error would occur even if a linearizer (Section 4.2) was being used and, generally, if the turbulence intensity is high enough to necessitate the use of a linearizer it is also high enough to necessitate some sort of correction for the effects of nondirectionality of the hot wire. The correction usually necessitates a knowledge of the triple products like $\overline{u^2v}$ or $\overline{v^3}$. Measurements of the triple products have been made in various flows (e.g. refs. 5, 38, 39): they are usually small near a solid surface but may be quite large near the edges of a turbulent flow. In the latter case, however, the turbulence intensity itself is small and the correction is not so important.

5.3.2. Flow with a Velocity Component Parallel to the Wire Axis (Fig. 35)

According to the equations of motion, the flow over an infinitely long wire in the plane normal to its axis should be independent of the velocity along the axis of the wire. The heat transfer, on the other hand, depends upon both the axial and the normal components of flow (consider, for

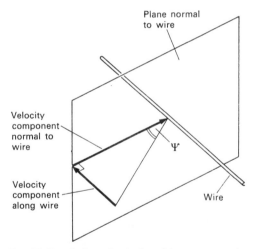

Plane normal
to wire

Velocity
component
normal to
wire

Ψ

Velocity
component
along wire

Wire

FIG. 35. Resolution of velocity with respect to wire.

instance, the case of flow entirely along the axis of the wire), but to a first approximation it is independent of the axial component of velocity. This "cosine law" for the effective cooling velocity U_{eff} as a function of the angle ψ between the flow direction and the plane normal to the wire axis can be used as the basis for empirical correlations of directional sensitivity: correlations include

(1) $U_{eff} = U \cos \psi$, the "cosine law";

(2) $U_{eff} = U(\cos^2 \psi + k^2 \sin^2 \psi)^{\frac{1}{2}}$, where, according to Champagne et al.,[40] k falls from 0·2 at $l/d = 200$ to zero at $l/d = 600$, the range of validity being $25 < \psi < 60$ deg;

(3) $U_{eff} = U(1 - k(1 - \cos^{\frac{1}{2}} \psi))^2$ where, according to Friehe and Schwartz[41], $k \approx 1 - 2600 \ (d/l)^2$ for hot wires and $1 - 2\cdot 2d/l$ for cylindrical hot films, the range of validity being $0 < \psi < 60$ deg.

The empirical constants that appear in these expressions depend on the length to diameter ratio of the wire because the non-uniform distribution of temperature along the length of the wire affects the directional response (the distribution along a yawed wire is asymmetrical). The latest work by Friehe and Schwartz seems to confirm that the cosine law is quite accurate for infinitely long wires, at least in the range $0 < \psi < 60$ deg. My own view, again based on experience, is that probes for directional measurements or for measurement of the lateral velocity fluctuations should be calibrated individually because of the effects of finite wire length and prong interference on the aerodynamic performance, and the difficulty of accurately setting or measuring the wire angle (the accuracy needed in this operation can be inferred from noting that the tangent of the wire angle, which is roughly what appears in the ratio of v component sensitivity to u component sensitivity, changes from 1·0 at 45° to 1·1 at 48°, so that extreme care in manufacture or measurement is needed to produce probes whose wire angle is known to within 1 or 2 per cent). Empirical determination of directional sensitivity is quite straightforward, providing of course that a stream of known direction is available and that the probe can be yawed through an angle, $\Delta\psi$ say, of about $\pm 5°$ from its null position. A simple device for ensuring accurate yaw angles is shown in Fig. 36. The best way of recording the results is to pretend that the cosine law is accurate and to work out an "effective" value of $\tan \psi$, that is, the ratio of the v component sensitivity to the u component sensitivity. ψ_{eff}, being to a first approximation the actual wire angle, will usually be independent of changes in the wire calibration caused by variations of fluid temperature, dust deposition and so on. If the wires are long enough for the variations with speed of the asymmetry of the temperature distribution to be unimportant, ψ_{eff} will also be very nearly independent of speed. At very low Reynolds numbers, the thickness of the boundary layers on the wire supports may vary considerably with speed and alter ψ_{eff}. An alternative is to measure the actual wire angle and fit one of the formulae shown above instead of the simple

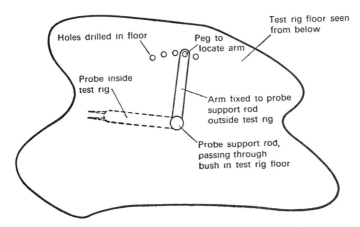

FIG. 36. Device for calibration at known yaw angles.

cosine law, that is, to find an apparent k instead of an apparent ψ. An experimental yaw calibration, of course, automatically includes an allowance for prong interference which will be valid for the larger turbulent eddies. A good indication of the cumulative effect of prong interference and other causes of departure from the cosine law can be obtained by plotting the wire voltage E against $\Delta\psi$ in the form

$$\left[\frac{E^2 - E_0^2}{(E^2 - E_0^2)_{\Delta\psi=0}}\right]^{\frac{1}{0\cdot45}} - \cos\Delta\psi \equiv -\tan\psi_{\text{eff}}\sin\Delta\psi$$

against $\sin\Delta\psi$ where E_0 is the calibration intercept (Fig. 34) at $\Delta\psi = 0$. Any marked deviation from a straight line on this plot (Fig. 37) indicates that something is wrong with the probe. If yaw calibration is not feasible care should be taken to ensure that the probe support interference is fairly small.

5.4. Contamination of Probes

Dirt in the fluid stream can cause large changes in probe calibration and the only cures are to filter the fluid, check the calibrations, clean the probes, and protect the probes from the fluid stream whenever possible. A change in probe calibration without a change in cold resistance is usually attributable to dirt. A layer of dirt on the wire will alter the frequency response as well as the mean calibration: it is a mistake to allow the dirt coating to reach an equilibrium state so that the mean calibration ceases to change, because such a dirty wire would have a very

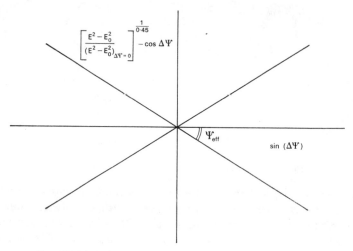

Fig. 37. Plot of yaw calibration to find effective wire angle.

poor frequency response. Dirt and slime deposition is an even more difficult problem in liquids than in gases though the effects on frequency response may not be so serious in practice because frequencies are usually lower. If the fluid is so dirty that the probes are actually broken by particle impact, it is likely that the change of calibration due to the deposition of smaller particles would be so rapid as to make accurate measurements almost impossible anyway. The smaller the sensing element, the sooner the thickness of the deposit will reach an appreciable fraction of its diameter. The deposit is usually confined to the upstream side of the probe near the stagnation line: for this reason, hot films on the *sides* of glass wedges or cones are frequently used for measurements in dirty fluids (Fig. 32).

It is difficult to remove dirt from a fluid stream effectively because no filter of reasonable pressure drop will remove, in a single transit of the fluid, all the very small particles that contaminate a hot wire. Filtering is easier in a recirculating test rig and it is worth noting that an open-circuit wind tunnel in a closed room is, in effect, a recirculating test rig. Electrostatic dust precipitators can easily be fitted in a by-pass loop in a wind tunnel or, if the speed is low enough, in the settling chamber of the tunnel itself, but mechanical filters must be used in liquid flows. Application of grease or one of the commercial "dust-attracting" compounds to the floor of the test rig may help to remove the larger particles which, although they would not rise into the fluid stream to affect the wire directly, are likely to disintegrate after continued impacts with the floor into smaller particles that will. Measurements near the floor are particularly likely to be affected by dirt in the fluid. In water, organic deposits may be important but as it is now generally realized that at least some types of algae may produce non-Newtonian effects (Section 3.10) and spurious drag reduction in turbulent flows in ship tanks and water test rigs, it is usual to guard against their appearance by chlorinating or otherwise treating the water. In water or other electrolytes, it is highly desirable that the probes should be electrically insulated from the liquid in order to prevent electrolytic or chemical action. Hot film probes with sputtered quartz coatings about 1 micron thick are commercially available: thin wires are difficult to coat, but cylindrical films are an acceptable substitute. These very thin quartz coatings will break down under quite small potential differences so that it is advisable to earth the "earthy" side of the probe to the liquid. Lacquer and polyurethane paint have been used on larger wires. Air bubble formation on heated probes in water can also be prevented by de-aeration, using a cavitating nozzle, but air or vapour bubble formation usually sets an upper limit to the probe temperature for water measurements[42]. Bubbles can sometimes be dislodged by squirting a high-speed jet of water at the probe just before taking a reading.

The best way of cleaning a probe is to immerse it in an ultrasonic solvent bath. The large baths used for degreasing machine components are not generally suitable and the small variety used by watchmakers and electronic engineers is preferable. Even these, when operated at full power, may cause changes in probe cold resistance during the first few seconds after immersion for the first time. This is particularly noticeable for probes made by copper-plating tungsten wire and subsequently etching it in an acid jet: the etched edges of the copper are generally rather ragged and are probably polished by the action of the ultrasonic bath. Further cleaning seems to have no effect either on the cold resistance or the calibration. The rate of industrial dirt deposition on a hot wire in atmospheric air is significantly less than the deposition on a cold wire: wires used for resistance thermometer measurements can be partially cleaned by heating them to a typical "hot-wire" temperature. However, it may be deduced from the performance of self-cleaning domestic ovens that a very high temperature (about 500°C) is needed to burn off all grease and oil deposits. It may be possible to clean probes used in water by heating them sufficiently strongly for local boiling to occur but care should be taken that this does not damage the probe.

When one is otherwise confident about the performance of the hot-wire system, dirt is a problem one can live with: but when one is faced with other causes of uncertainty, such as large temperature changes in the flow, or when one is trying to develop or check new techniques or apparatus, it is worth going to a great deal of trouble to keep the probes and the fluid clean.

5.5. Probe Design and Manufacture

The three most common configurations for hot-wire probes (Fig. 38) are:

(1) a single wire normal to the stream, used for measuring the u component fluctuation ("U" probe);

(2) a pair of wires arranged in an "X" at approximately $\pm 45°$ to the flow direction: ideally, one wire of an "X" probe in the xy-plane responds to $u + v$ and the other to $u - v$, according to the simple cosine law, so that the difference between the two wire signals is $2v$: in practice, the wires cannot be made electrically and aerodynamically identical and the u component and v component sensitivities will be different for the two wires. The angle is not critical: angles between 30° and 60° give the best results. The planes of the two wires should be separated by not less than half the wire length, so that the heated wake of one wire does not affect the other;

(3) the slant-wire probe, which looks like half an "X" probe and responds (roughly) to $u + v$. If rotated 180° about the probe axis, it will respond to $u - v$ and the difference between the mean squares of the fluctuating signals in these two positions is $4\overline{uv}$. When used in this way, the slant-wire probe simulates an "X" probe with two perfectly matched wires, but, of course, it cannot be used for instantaneous measurement of the v component.

Sometimes a pair of wires is arranged in a "V" rather than an "X" configuration; this type of probe is easier to manufacture but its spatial resolution is not as good as that of an "X" probe. More complicated arrays of wires are used for special purposes: for instance, if both velocity and temperature fluctuations occur in the stream a cold temperature-measuring wire can be added either to a normal-wire probe or an "X" probe (Chapter 7). Arrays of three mutually perpendicular wires have been used for flow direction measurements but the effects of prong interference and departures from the cosine law makes their calibration rather uncertain. Probes with four wires have been used for measuring one or other component of the fluctuating vorticity; it can be seen that the probe shown in Fig. 38 will have an output proportional to $\partial w/\partial y - \partial v/\partial z$ which is the x component of vorticity. For simplicity, Fig. 38 shows only hot-*wire* probes: some at least of these configurations can also be set up with hot-film probes, although the directional sensitivity of films is often poor because the effective heat transfer area has a much lower aspect ratio than the film itself, because of conduction through the substrate.

A wide variety of hot wire and hot film probes is available commercially but, although these are of very high quality, they are rather expensive and many users may prefer to make their own (with or without working out what this really costs). One should always be prepared to discard probes that are giving erratic readings for no discoverable reason, and whether the experimenter is more willing to do this with a probe that he has bought at a high price or one that he has made himself is an interesting psychological question. The design of the simpler hot-wire probes at least is quite straightforward: the prongs and body must be slender enough not to interfere with the flow round the wire (see ref. 37) but strong enough not to allow the wire to vibrate. The simplest way of assessing interference at the design stage is to ask "would I believe the

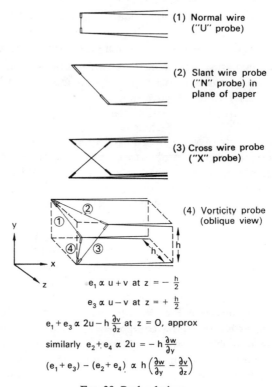

(1) Normal wire
("U" probe)

(2) Slant wire probe
("N" probe) in
plane of paper

(3) Cross wire probe
("X" probe)

(4) Vorticity probe
(oblique view)

$e_1 \propto u+v$ at $z = -\frac{h}{2}$

$e_3 \propto u-v$ at $z = +\frac{h}{2}$

$e_1 + e_3 \propto 2u - h\frac{\partial v}{\partial z}$ at $z = 0$, approx

similarly $e_2 + e_4 \propto 2u = -h\frac{\partial w}{\partial y}$

$(e_1 + e_3) - (e_2 + e_4) \propto h\left(\frac{\partial w}{\partial y} - \frac{\partial v}{\partial z}\right)$

FIG. 38. Probe designs.

reading of an infinitesimal anemometer placed anywhere along the wire?" Interference is automatically taken into account when the probe is calibrated in a mean flow, but the response to small eddies (wavelengths not much larger than the wire length) will not be affected so much by prong and body interference. This worsens the effects of finite spatial resolution (Section 5·6) in u component measurements but opposes these effects in v or w component measurements, because the effect of prong interference is to increase the velocity over the wire but to decrease the effective yaw angle. Vibration will obviously cause spurious velocity fluctuations, but it may also cause "strain gauging" in the wire leading to further errors. Vibration in the probe or in the test rig itself usually occurs at discrete frequencies and can easily be detected by doing a spectrum analysis—a worth-while precaution whenever a new probe support arrangement is being used. The usual technique for making general-purpose probes is to imbed sewing needles, with electrical connecting leads soldered on, in a small block of plastic (epoxy resin or similar). At low speeds there is no real need for the electrical leads to be enclosed within the probe body but long pieces of dangling wire may change their position and alter the inductance of the probe appreciably, thus affecting the frequency response, although this effect is not too critical at low speeds and is minimized by twisting the leads together.

There are two main techniques for attaching the anemometer wire itself to the prongs. Commercial probes are usually made by *welding*. It is not possible to make a true weld with tungsten because of its high melting point: what actually happens is that only the material of the prong is melted, making a rather poor "soldered" joint. It is, therefore, advisable to plate a thin coating of copper or platinum on to wire before attempting to weld it. Platinum-plated tungsten wire is commercially available and so is suitable manipulating and welding apparatus. Wollaston wire with a platinum core is usually *soft soldered* to the prongs and the silver coating then etched away over the central portion of the wire. The simplest method of etching is to use a capillary jet of acid from a burette held above the wire: jet diameters as small as 0·5 mm are easily obtained. It is best to use weak acid (say 10 per cent nitric acid) and speed up the process by applying a positive potential of about 6 volts to the wire, completing the circuit by a platinum electrode in the burette: the advantage of this system is that acid splashes are less harmful to the probe and the surroundings than if stronger acid were used; in particular, the two wires of an "X" probe can be etched independently by applying the voltage to each in turn. It would, of course, be possible to weld or solder bare platinum wire but Wollaston wire is much easier to manipulate and another advantage is that, by traversing the wire with respect to the etching jet, the length of the etched portion can be quite closely controlled and the active portion of the wire kept outside the region of prong interference by leaving unetched "sleeves" (Fig. 32) at the ends. Soft soldering, using a small electrically heated iron of the type used by electronic wiremen, is simple and a manipulator is not really needed, even for the manufacture of "X" probes. However, the welding technique is almost as simple once the necessary apparatus has been set up and is much to be preferred if one is trying to set the wire at an accurate angle to the probe axis. The "shelf life" of welded probes is considerably better than that of soft-soldered probes although this is not necessarily an advantage because it tempts one to carry on using probes for a very long time with the possibility of irreversible dirt contamination. Although welding tends to go with tungsten and soldering with platinum-core Wollaston wire, this is not an unalterable rule; a convenient method of handling tungsten is to copper-plate it up to several times its original diameter and treat it like Wollaston wire. A welded tungsten wire can have its ends plated after assembly so as to leave only a central portion of "active" wire.

It is best to leave the wire very slightly slack or bent: a straight pretensioned wire may produce large "strain gauge" signals, particularly if the prongs vibrate. The desired effect is obtained in soft soldering by letting the wire hang free while making the second joint and, with care, may also be achieved when welding.

Hot films are also available commercially and their greater durability makes the price less important so that few people will bother to make their own unless special shapes are required. Sputtering is the best (and most expensive) technique: the alternative of applying platinum paint and then firing in a furnace seems to give less reliable results. In both cases, joining the thick connecting leads to the thin film presents a problem: again, sputtering seems to be the best technique.

5.6. Spatial Resolution

The theory of the response of a finite measuring instrument to a general random field is very complicated. Fortunately, simplifications can be made in the case of hot-wire measurements of turbulence, taking advantage of the simple shape of the wire and the universal nature of the small-scale motion (Section 2.5). Since the small-scale (isotropic) motion at a given wave number k depends only on ν and ε, the ratio of the spectral density measured by a fairly short wire of length l to that measured by a *very* short wire of the same sensitivity is a universal function of $k\eta$ and l/η, where $\eta \equiv (\nu^3/\varepsilon)^{\frac{1}{4}}$ is the Kolmogorov length scale, and of the angle between the wire and the stream. In the inertial subrange, where ν does not influence the motion, the ratio must be a universal function of kl only. If the wire is so long as to attenuate the anisotropic energy-containing eddies, a prohibitive amount of information about the spectrum is needed, and virtually the only course open in practice is to use the "isotropic" correction formula and hope that it is not too much in error.

The universal function $f(k_1\eta, l/\eta)$ defined by

$$\frac{\Phi_{\text{measured}}}{\Phi_{\text{true}}} = f(k_1\eta, l/\eta)$$

for a wire normal to the stream, depends on the universal spectrum, about which there is still some uncertainty: Wyngaard[43] has calculated f using an analytical form of the wave-number-magnitude spectrum $E(k)$ (see Section 2.6):

$$E(k) = \alpha\varepsilon^{\frac{2}{3}} k^{-\frac{5}{3}} \exp\left[-\frac{3}{2}\alpha(k\eta)^{\frac{4}{3}}\right]$$

with $\alpha = 1\cdot7$, and this should be adequate for all hot-wire length corrections. For $l/\eta > 3$, f falls to $0\cdot5$ at k_1l of the order of 5 (wavelength $\simeq l$). Wyngaard also discusses the length corrections for a cross-wire probe with a finite distance between the two wires: in this case there is "crosstalk"—for instance, the measured u component spectrum contains large contributions from the v component.

This theory ignores the effect of heat conduction to the supports and assumes that the wire sensitivity is uniform along its length: in practice the effective length can be taken as $l - 2l_c$.

5.7. Frequency Response

In this section we consider only an infinitely long wire, the effect of heat conduction to the supports on the frequency response being discussed in Section 5.1.1. The frequency response of a hot film, other than the cylindrical type, is almost entirely determined by heat conduction to the supports, the thermal capacity of the film itself being negligible in comparison.

The response of a wire of mass m_w and specific heat c_w is governed by the first-order differential equation

$$m_w c_w \frac{dT_w}{dt} = I^2 R_a(1 + \alpha(T_w - T_f)) - \pi k_f l(T_w - T_f)\,\text{Nu}$$

whose solution includes the term $\exp(-t/M)$ where

$$M = \frac{m_w c_w}{I^2 R_a \alpha - \pi k_f l \, \mathrm{Nu}} = \frac{m_w c_w}{\pi k_f l \, \mathrm{Nu}} \cdot \frac{R_w}{R_a} = \frac{\varrho_w c_w d^2}{k_f \, \mathrm{Nu}} \cdot \frac{R_w}{R_a}.$$

The wire behaves, in fact, like a low-pass filter of time constant M (Fig. 42). The mean-square response to a small sinusoidal oscillation of Nu with frequency ω and constant amplitude varies as

$$\frac{1}{1 + \omega^2 M^2}$$

(the mean-square frequency response is the Fourier transform of the step response, just as a frequency spectrum is the transform of the auto-correlation). If the fluctuations in T_w are large, the response is complicated by changes in fluid properties and in the effective value of α.

The time constant M of our typical wire is about 0·6 msec (0·6 × 10⁻³ sec). Since M depends strongly upon the wire diameter it is not usually possible to calculate it with sufficient accuracy (also, the physical properties of very fine wires are not always the same as those of the same material in bulk). Therefore, it is usual to measure it directly, generally by superposing a small oscillation on the heating current I rather than on Nu. About the only convenient way of oscillating the flow is to superimpose sound waves.

Clearly, our only real control over M lies in choosing the wire dia-meter: usually we make the wire diameter as small as possible, being limited by mechanical strength or availability. Small reductions in time constant, at the expense of sensitivity, can be obtained by reducing R_w/R_a (since this also reduces I, the overall frequency response of a constant-temperature system may be worsened: the overall time constant is $M/$(amplifier gain) and the denominator generally increases with I). M is smaller in water for given Nu because k_f is larger but frequency response of hot wires is not usually so critical in liquid flows because speeds are usually lower.

In low-speed flows, performance is usually limited by spatial resolution rather than the frequency response of the anemometer system. Since the small-scale turbulence is nearly isotropic, the frequency at which the mean-square response of the anemometer has fallen to half its nominal value (the "−3 dB point") need not be much larger than $k_1 U$, where k_1 is the wave number at which the spatial resolution factor is one-half.

EXAMPLE. For our typical wire, operated in a boundary layer in air, at a speed of $U = 30$ m s^{-1} at $y = 0\cdot5$ cm from the surface, Wyngaard's results give $k_1 l \simeq 3$ (taking $\varepsilon = u_\tau^3/0\cdot41y$) so that $k_1 U/2\pi = 15{,}000$ Hz. In this example, l/y is as large as $0\cdot2$ and the dissipation would be underestimated by roughly a factor of 3 unless corrections for poor spatial resolution were made: $\overline{u^2}$ would be underestimated by very roughly 10 per cent, or 5 per cent on $\sqrt{\overline{u^2}}$, which is an unacceptably large consistent error for many purposes, even if random errors of this magnitude may have to be accepted. Note that poor spatial resolution is *always* more important than heat conduction to the surface, although the latter must be taken into account when measuring *mean* velocity in the viscous sub-layer.

CHAPTER 6

Analysis of Fluctuating Signals

"Two added to one—if that could but be done",
It said, "with one's fingers and thumbs!"
Recollecting with tears, how, in earlier years,
It had taken no pains with its sums.

To FORM the statistical mean values outlined in Chapter 2, we have to add, subtract and multiply the fluctuating outputs of the measuring instruments, and sometimes delay them in time or filter them in frequency bands. These operations can be done by analogue or digital techniques and apparatus for either is commercially available. The older textbooks tend to reflect the position of a few years ago when electrical analogue computing apparatus had a standard bandwidth of 100 Hz and digital computers were too slow and too small for analysis of wide-band digital records to be practicable. More recently, reliable analogue equipment with ample bandwidth has become available, some of it specially designed for use with hot-wire anemometers, and there is no difficulty in obtaining apparatus to perform all the statistical operations mentioned in this book. Digital recording systems are also in common use (Section 6.6) but have the great disadvantage that results are not usually available in real time. This book is intended as an introduction for lone workers rather than for established groups, and it is the latter who are more likely to have access to digital systems or on-line computers. Therefore, most of the discussion in this chapter refers to analogue devices: a more detailed introduction is given in ref. 44.

6.1. Analogue Computing Elements

An experimenter can, if he wishes, treat his electronic apparatus as a collection of "black boxes" which will perform good approximations

134

to certain mathematical operations. In this book we encourage a cautious version of this attitude, necessary to ensure that the approximations really *are* good. To start with, the experimenter must have some idea of principles of operation before he can choose suitable commercial apparatus.

6.1.1. *Addition and Subtraction*

Addition of two voltage signals can be performed by converting both into *currents* and applying them to a resistor, as shown in Fig. 39.

The arrowhead is the standard symbol for an amplifier of very large voltage amplification or "open-loop gain", A, say, unless otherwise stated: analogue computing circuits almost always provide "feedback"

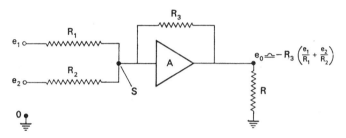

The amplifier supplies the load resistor R with
whatever current is required to maintain e_0

FIG. 39. Summing amplifier.

of the output to the input ("closed-loop" operation) so that very large voltages do not appear. For instance, the voltage at the "summing junction" S in Fig. 39 is e_0/A, which is effectively zero. Therefore, the sum of the input currents is $e_1/R_1 + e_2/R_2$, which must flow through the "feedback resistor" R_3 since the input resistance of the amplifier is arranged to be very high: it follows that

$$e_0 = -R_3\left(\frac{e_1}{R_1} + \frac{e_2}{R_2}\right) = -(e_1 + e_2)R_3/R_1 \quad \text{if} \quad R_2 = R_1,$$

again neglecting the voltage e_0/A at S and assuming that the inputs are supplied from low-resistance sources capable of providing any required current.

A special case of this circuit is the inverter or "minus one amplifier" in which $e_2 = 0$, $R_1 = R_3$ and $e_0 = -e_1$. Clearly, subtraction can be performed by supplying one input to the addition circuit through an inverter, but more compact "differential" amplifiers are normally used.

6.1.2. Differentiation and Integration

The resistances in the above circuits may be replaced by inductances or—more usually—capacitances. For instance, the circuit shown in Fig. 40 acts as an integrator: if

$$e_1 = a \sin \omega t, \quad \text{then} \quad e_0 = -\frac{a}{\omega} \cdot \frac{1}{RC} \cdot \cos \omega t,$$

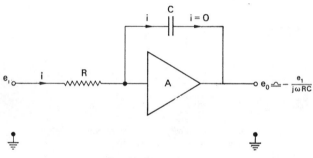

FIG. 40. Integrator.

a result most easily arrived at by following the usual complex-number notation for a.c. circuits and regarding C as a resistance-plus-phase-shift (impedance) equal to $-1/j\omega C$ where $j = \sqrt{-1}$. Alternatively, if the input is zero for $t < 0$ and equal to e_1 for $t > 0$ then the voltage across the capacitor, equal to e_0 if the amplifier gain is large, is given by

$$C \frac{de_0}{dt} = i = \frac{e_1}{R},$$

so that

$$e_0 = \int \frac{e_1}{RC} dt,$$

a result that is true in general and not only for constant e_1.

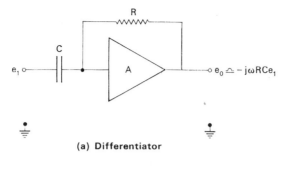

(a) Differentiator

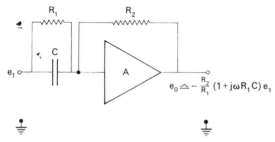

(b) Hot wire compensation circuit

FIG. 41. Differentiation circuits.

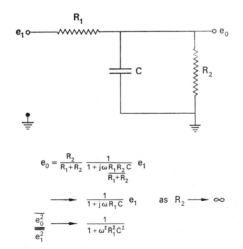

FIG. 42. Low-pass filter: without the capacitor and its lead, this is the simplest form of attenuator.

Similarly, the circuits shown in Figs. 41(a) and 41(b) respectively differentiate the signal and generate an output inverse to that of a simple low-pass filter of the type shown in Fig. 42. As seen in Section 5.7, a simple low-pass filter is an accurate analogue of the response of a hot wire, so that the circuit of Fig. 41(b) is a compensation circuit for a constant-current anemometer.

Neither the differentiation circuit nor the compensator will really work up to infinite frequency, because the amplifier does not have an infinite voltage gain: an accurate expression for the output of the differentiation circuit is

$$e_0 = \frac{e_1 A . j\omega RC}{1 - A + j\omega RC}$$

which tends to $e_1 A$ as ω tends to infinity and has an amplitude of $1/\sqrt{2}$ of the ideal value when $\omega \doteqdot A/RC$: this is also the approximate "-3 dB point" of the compensator. Note that this upper frequency limit has nothing to do with the frequency response of the amplifier itself (though obviously the latter will dominate if it is worse than that of the idealized circuit).

6.1.3. *Squaring, Multiplication and Linearization*

In the simple circuits described above, the amplifiers (valve or transistor) were supposed to amplify the signal without distortion—that is, the gain de_0/de_1 was assumed constant. In practice, the gain depends on (among other things) the instantaneous input, namely the grid voltage in the case of a valve, the base voltage in the case of a transistor, or the potential difference in the case of a diode. If $de_0/de_1 = A + Be_1 = B(e_1 + A/B)$ where A and B are constant, then $e_0 = B(e_1 + A/B)^2/2 + C$ where C is another constant, so if we subtract a constant voltage A/B from a signal before it is fed to the amplifier the output will be proportional to the *square* of the original signal. Obviously, the smaller A/B and C are, the less the results will be affected by amplifier drift; also, the wider the range over which $de_0/de_1 = A + Be_1$, the larger the ratio of the acceptable signal range to A/B. In practice, squaring circuits use valves or transistors biased fairly near the point at which the device switches off completely, so that the gain varies rapidly with input: care is needed to choose a device for which A and B are independent of e_1

and remain constant with time, so that highly stable power supplies are needed and the non-linear device is often mounted in a miniature temperature-controlled oven, using a thermal expansion of some sort to switch a heating element on and off. More than one non-linear device can be linked together: the biased-diode function generator is widely used to produce a piecewise-linear approximation to a given function, for instance to linearize the output of an anemometer. A diode is ideally a device that has an infinite resistance below a certain applied voltage (say zero) and a small, constant resistance above this voltage. If a number are arranged in parallel, with different, negative bias voltages applied in series with each, the total resistance will decrease in steps as the input voltage is increased and the net voltages across the diodes become positive one by one, so that the total current flowing in a common output resistor will be a piecewise-linear function of the input voltage. In practice, the bias voltages are applied by a chain of potential-dividing resistances and not by individual voltage soures, and the circuit arrangement is more difficult to understand. The effective resistance of real diodes changes gradually from a very large value to a small value as the bias voltage is increased, so that the characteristic of the function generator is smoothed out: this is an advantage in practice, and can be exploited by using transistors, connected as diodes, instead of actual diodes, thus reducing the number of stages needed to give an acceptably smooth characteristic. It is obvious that biased-diode function generators work best for *large* signals covering many segments of the characteristic: if a biased-diode hot-wire linearizer were used in measurements of weak turbulence over a range of speeds the apparent turbulence intensity would increase in steps as the d.c. voltage passed from one segment to the next.

Given a squaring circuit, multiplication can be performed, instantaneously or in the mean, by the quarter-squares technique:

$$\tfrac{1}{4}\{(u+v)^2 - (u-v)^2\} = uv.$$

Alternatively, logarithmic function generators can be used: $\log uv = \log u + \log v$. There are, however, many direct methods of multiplication, falling into three main groups, of which examples are:

1. *Transistor (or valve) characteristic.* The current flowing through one terminal of a transistor (say the collector) depends on the voltages at the other *two* terminals (base and emitter) and is therefore proportional

to their product (possibly after the application of bias voltages as in the case of the squaring circuit: the same cautionary remarks apply). Practical multipliers use rather more sophisticated arrangements of semi-conductors.

2. *Time division*. One input signal is used to modulate the mark/space ratio of a high-frequency oscillator which switches the other input signal on and off: successive output pulses thus have a width proportional to the first signal and a height proportional to the second, and the instantaneous product is recovered by passing the output through a low-pass filter whose cutoff frequency lies between the signal frequency and the oscillator frequency.

3. *Hall effect*. If a magnetic field B and a current i are applied to crystals of certain materials, a potential difference proportional to the vector product $i \wedge B$ is produced: in principle the frequency response is very high but is limited in practice by that of the coil used to produce the magnetic field (phase shifts are particularly important).

Type 1 seems to be much the most popular in commercial apparatus, but, at the time of writing, the best accuracy is obtained with quarter-squares circuits: this is probably a law of nature, since the quarter-squares technique makes less severe demands on transistor performance, but the quarter-squares arrangement needs three addition or subtraction circuits, which may themselves introduce inaccuracies. Either a quarter-squares arrangement or the direct (type 1) multiplier needs special or duplicated circuits to accept positive and negative values of both signals (called "four quadrant" operation).

6.1.4. *Frequency Analysis* (Fig. 50)

In the past, measurements of frequency spectra have usually been preferred to measurements of autocorrelations because of the inconvenience of mechanical time-delay apparatus such as acoustic delay lines or tape recorders with movable heads. The appearance of digital or semi-digital time-delay equipment has altered the position because such equipment is rather easier to use, in principle at least, than a frequency analyser. However, frequency analysis is straightforward providing that the principles are understood. The mathematics of frequency spectra were discussed in Section 2.5: in this section we discuss the choice of equipment and its use.

There are three common types of frequency analyser (or "spectrum analyser" or "waveform analyser" or "spectrometer"):

(1) A set of passive filters, that is a set of inductance–capacitance band-pass filters, one for each of the desired frequencies: generally the band-width of each filter is made directly proportional to the centre frequency, which makes design rather easier and is also convenient in practice if smooth spectra covering a wide frequency range are to be measured. Filters with a bandwidth of one-third of an octave ($2^{\frac{1}{3}}$ times the centre frequency) are commonly used in acoustics, but such wide-band filters are not suitable for spectra with sharp peaks.

(2) A Wien bridge feedback circuit, in which a feedback amplifier is used to reject the input at all but a selected band of frequencies. Again, the bandwidth is a constant percentage of the centre frequency, but the advantage is that the centre frequency can be continuously adjusted by changing a capacitor in the feedback circuit. A disadvantage is that the maximum attenuation of the input at frequencies far from the chosen centre frequency is equal to the reciprocal of the amplifier gain, which is severely limited in some commercial equipment of this type, whereas the maximum "off-tune rejection" of the passive filter network is limited only by stray induction between components and can be very large: this point is important when making measurements at low centre frequency (and therefore small bandwidth) if the unfiltered signal covers a wide frequency range.

(3) A "heterodyne" circuit, in which the input signal is multiplied by a sinusoidal signal at the required centre frequency, ω_0 say: the resulting signal has components at very low frequency, arising from the components of the input signal near the centre frequency, and if the resulting signal is fed through a *low-pass* filter of cutoff frequency $\Delta\omega/2$, the mean square of the output will be proportional to that which would be obtained by feeding the original signal through a *band-pass* filter of centre frequency ω_0 and bandwidth $\Delta\omega$. In practice, accurate multiplication is not required and rather simpler "mixing" circuits are used. Also, commercial apparatus usually employs two sinusoidal signals, 90° out of phase, which is necessary if the signal to be analysed contains a discrete frequency whose phase difference from the sinusoidal signal is constant but unknown. If the discrete-frequency part of the signal to be analysed is

$e_1 \sin(\omega t - \phi)$ we have

$$e_1 \sin(\omega t - \phi) \sin \omega_0 t = \frac{e_1}{2} \{\cos[(\omega - \omega_0)t - \phi] - \cos[(\omega + \omega_0)t - \phi]\},$$

$$e_1 \sin(\omega t - \phi) \cos \omega_0 t = \frac{e_1}{2} \{\sin[(\omega - \omega_0)t - \phi] + \sin[(\omega + \omega_0)t - \phi]\}$$

and the sum of the squares of the low-frequency components (those at frequency $\omega - \omega_0$) is $e_1^2/4$, independent of ϕ. If ϕ varies randomly, as it does in typical turbulence signals, it does not affect the mean square output. Heterodyne analysers necessarily have fixed bandwidths, but commercial instruments usually offer a choice, obtainable by changing output filters.

On balance, sets of passive one-third octave filters are the most suitable for turbulence work but need supplementing by one of the other two types if vortex shedding or other discrete-frequency phenomena are being studied. The error involved in using such wide-band filters with rapidly changing spectra is surprisingly small, and negligible in nearly all cases.

Presentation of the results of spectrum measurements sometimes causes difficulty, mainly because of carelessness in making the results dimensionless. The best course is to plot spectral density ϕ divided by the overall mean square value, $\overline{u^2}$, so that

$$\int_0^\infty \frac{\phi}{\overline{u^2}}\, d\omega = 1.$$

Spectral density per Hertz is simply 2π times spectral density per radian/sec, so that either measure of frequency can be chosen: it is best to make the frequency dimensionless, also, and since one is trying to approximate a wave number spectrum the obvious choice is $\omega l/U$ where U is the mean velocity at the measurement point, l a typical length scale and ω the radian frequency: this is an approximation to kl. Therefore the spectral density per unit $\omega l/U$, made dimensionless by $\overline{u^2}$, is obtained as

$$\left. \frac{\text{mean square filter output}}{\text{overall mean square value}} \right/ (\Delta\omega\, l/U)$$

where $\Delta\omega$ is the filter bandwidth in radians/sec.

It is usual to plot turbulence spectra on log-log paper because of the wide range of spectral density and frequency to be covered: this does not give a good idea of the relative contribution of each frequency range to the overall mean square, and an alternative is to plot $\omega\phi$ against $\log \omega$. Since $\int \omega\phi d (\log \omega) = \int \phi \, d\omega$, equal areas under different portions of the graph contribute equally to $\overline{u^2}$. If $\Delta\omega/\omega_0 = $ constant, $\omega\phi$ is directly proportional to the mean-square filter output. In acoustic work, ϕ or $\omega\phi$ is usually measured and plotted in decibels, which amounts to taking the logarithm.

6.2. Input and Output Impedance, and Frequency Response

"Impedance" means "resistance and/or reactance": usually the relation between input voltage and input current of an electronic box approximates that of a large resistance with a small capacitance in parallel with it; the apparent resistance and capacitance may vary with operating conditions. For instance, in the circuit of Fig. 43 the resistive part of the input impedance is roughly $\alpha_{fb}R_e/(1 - \alpha_{fb})$ which depends on the "common-base current gain" α_{fb} but is much larger than R_e. Both α_{fb} and the capacitance C_{eb} depend on the collector current. The output impedance, or "source resistance", $de_0/di_0 \equiv Z_0$ is much *smaller* than R_e. If e_0 feeds an impedance Z the output voltage will be $Z/(Z + Z_0)$ times the "open circuit" value. In practice, Z will be the combined

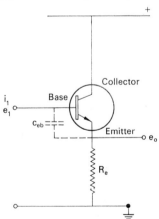

FIG. 43. NPN emitter follower.

impedance of the connecting cable and the input impedance of the next box in the chain, and this impedance must be made large compared with Z_0 if the output is not to be affected. If Z and Z_0 are pure resistances, the attenuation of the output is the same at all frequencies and could be allowed for in calibration (providing that we remembered to recalibrate if we add another impedance in parallel with Z, say by connecting an oscilloscope to look at the output signal): it is usually possible to make Z_0 small enough and Z large enough (by using another emitter follower in the input to the second box) for boxes to be connected without appreciable attenuation—though a check, even if only by reference to the manufacturer's specification, is always advisable.

The impedance of the connecting cables can usually be neglected, at least for the lengths commonly used in the laboratory. Coaxial cable (in which the "earth" conductor is a sleeve, usually of wire braid, surrounding the "live" conductor and separated from it by a plastic dielectric material) should be used for all connections between units in order to prevent the live conductor acting as an aerial antenna and picking up unwanted signals from the electrical mains or from radio transmitters. The capacitance between the live and earth connectors is larger than for other types of cable (about 100 pF per metre of cable length) but this is outweighed by the other advantages. The resistance is low and the inductance between the conductors is nominally zero, so that the cable is closely equivalent to a capacitance across the output of the box that feeds it, and the cable capacitance and the output resistance combine to form a low-pass filter. If a -3 dB point at 100 kHz is required and the cable capacitance is 100 pF, the output resistance must be less than 16 kΩ: typical output impedances of valve amplifiers are 1 kΩ or less and transistor amplifiers have output impedances and order of magnitude less than this so that no difficulty should be experienced unles the *total* length of connecting cable in the chain exceeds some tens of metres. When only very low frequencies occur, as in liquid flows, even less care is needed and coaxial cable is not essential; however, pick-up from the electrical mains is likely to be a nuisance in early all experiments and may even affect the d.c. reading of certain types of digital voltmeter.

Apart from the effect of connecting cables, the frequency response of a chain of processing equipment is the *product* of the frequency responses of the individual units: if there are n units, each equivalent to a simple low-pass filter with an upper "$-3\,\mathrm{dB}$" point at a radian frequency ω_0, the $-3\,\mathrm{dB}$ point of the chain is at $\omega_0\sqrt{(2^n-1)}$ or $0\cdot5\omega_0$ if $n=3$ and $0\cdot2\omega_0$ if $n=10$. In practice, a given unit will itself be composed of several low-pass filters and will thus have a sharper cutoff, so that the above formula gives a pessimistic estimate. As well as producing an attentuation, imperfect frequency response also produces a phase shift, $\tan^{-1}\omega/\omega_0$ for a simple low-pass filter: this phase shift can be extremely large at the end of a long chain of equipment, but is immaterial in turbulence studies, providing that all signals suffer the same phase shift; because there is no correlation between the phases of turbulent fluctuations of different wave number ($\simeq$ frequency), the phase at a given wave number fluctuates randomly, and therefore meaningful statistical quantities are little affected by variation of phase shift with frequency.

The most spectacular example of a large phase shift occurs in frequency analysers which are basically band-pass filters: a single band-pass filter, consisting of an inductance and capacitance in series, has a 90-deg phase lead at frequencies much lower than the centre frequency and a 90-deg phase *lag* at very high frequencies. This does not affect the measured power spectrum at all, although difficulty may be experienced if two analysers are being used, as in direct measurements of the $\overline{uv}$ spectrum, because the phase shifts of the two are bound to be slightly different: the simplest course is to feed the same turbulence signal into the two analysers when set to the same frequency, measure the correlation coefficient of the two outputs, divide all measured cross-spectral densities at this frequency by the correlation coefficient and take the average of the two sets of measurements obtained by interchanging the filters. More strictly, one should deduce the average difference in phase angle between the two filters which, being small, is easier to deduce by feeding a signal u into one filter, a signal du/dt into the other and equating the correlation coefficient to the *sine* of the phase angle difference β: then if the true cross-spectral densities of $\overline{uv}$ and $\overline{u\,\partial v/\partial t}$ are Φ and Φ', the measured quantities are

$$\Phi\cos\beta+\Phi'\sin\beta,$$

$$\Phi'\cos\beta-\Phi\sin\beta$$

respectively. The "quadrature" spectrum Φ' (so called because there is a 90-deg phase shift between a signal and its time derivative) is not necessarily zero although $\overline{u\,\partial v/\partial t}+\overline{v\,\partial u/\partial t}=\partial\overline{uv}/\partial t=0$. However, it can be seen that since β changes sign when the filters are interchanged, the effect of Φ' can be made to cancel out as above when only Φ is wanted: the longer procedure is necessary if Φ' is itself wanted, as in some measurements relating to convection velocity.

In most correlation measurements, it is sufficient for duplicated units to be *nominally* identical (two of a kind from the same manufacturer): if the correlation between the signals obtained from different types of tranducer is required (say in pressure-velocity correlations) care is needed to match or measure the phase shifts of the two channels.

Low-frequency response and phase shift may be as important as high-frequency behaviour. However, the high-frequency response is limited by the irreducible parasitic capacitances and inductances in the circuits, whereas low-frequency response depends on the "decoupling" capacitors used to maintain different parts of a circuit at different d.c. voltages; these capacitors can be made as large as we like and it is usually safe to increase the capacitances in commercial units when necessary. A very good low-frequency response is a nuisance in practice because recovery from transient loads, occurring during switching or adjustment, is correspondingly slow. A direct-coupled circuit does not "recover" in this sense (Fig. 44) and so it is an advantage to use d.c. circuits where possible: d.c. amplifiers are very much easier to construct using transistors than using valves because n–p–n and p–n–p transistors can be used together, avoiding a continuous increase in d.c. bias from input to output, but care has to be taken that short-term or long-term drift does not cause errors or overloading of the circuits. When a.c. coupling must be used, the lower "−3 dB" points should be chosen in the flat part of the turbulence frequency spectrum so that the latter can be extrapolated to zero frequency.

6.3. Noise and Hum

Resistors, transistors or valves produce spontaneous electrical signals due to electron motion: the voltage fluctuations are closely Gaussian (see Glossary). Resistor noise has a power spectrum that is flat up to extremely high frequencies, whereas one of the most important components of transistor noise has a spectral density inversely proportional to frequency down to extremely *low* frequencies. Generally, the main source of noise is the input stage of a low-level amplifier: elsewhere in the processing apparatus, the signal level is (or should be) much higher

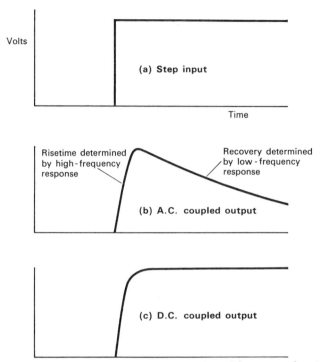

Volts

(a) Step input

Time

Risetime determined by high-frequency response

Recovery determined by low-frequency response

(b) A.C. coupled output

(c) D.C. coupled output

FIG. 44. Response of a.c. and d.c. coupled amplifiers to step input.

than any locally generated noise. Thus the only important source of noise in hot-wire anemometry is the anemometer amplifier itself. In commercial circuits, care is taken to use low-noise transistors and to keep resistances in the input stage as small as possible, and little more can be done. Low-pass filters can be used to cut off the noise generated at frequencies above those found in the turbulence signal but this does not affect the signal-to-noise ratio at a given frequency. The hot wire can be connected to its a.c. amplifier by a step-up transformer: ideally, a transformer of voltage ratio n is a noiseless amplifier of gain n but there are practical disadvantages.

Since the amplifier noise is uncorrelated with the signal, its mean-square value can be subtracted from the measured mean-square value of signal plus noise: this assumes that the noise can be measured independently, which is not easy if the noise varies with the hot-wire operating conditions. The noise of a constant-temperature anemometer tends to increase with flow speed because the frequency response increases: this effect is not noticeable if a suitable fixed filter is used. The noise at constant frequency decreases with increasing flow speed because the wire time constant decreases: in the case of a constant-current anemometer, the noise can be measured with the wire current switched off (but not with the wire disconnected from the amplifier: noise depends on the impedance of the source from which the amplifier is supplied).

Hum or mains pickup arises either from electromagnetic induction in connecting leads inside or outside a unit (particularly from so-called "earth loops" caused by finite contact resistance or lead resistance in earth connections, which often have complicated, multiply-connected paths) or from inadequate smoothing in the d.c. power supply (the mains frequency content of a power supply output is usually called "ripple" to distinguish it from pickup). Harmonics of the mains frequency may also occur: three-phase controllable rectifiers (mercury-arc or thyristor type) produce high-frequency transients with a repetition rate three times the mains frequency.

"Microphony" (vibration sensitivity) is important only in units containing valves or piezoelectric crystals (such as Hall effect multipliers): its appearance in other circuits implies a loose connection or an inadequately mounted component. The sensing element (hot wire, laser optical system) may itself be microphonic.

These sources of spurious signal, usually uncorrelated with the required turbulence signal, are the same in their effect as the free stream turbulence in the test rig, which is also (usually) uncorrelated with turbulence in shear layers: in most cases the two can be measured together, and a "free stream" value subtracted from all measured quantities. Indeed, it is rather difficult to distinguish test rig turbulence from pickup due to the test rig drive.

6.4. Averaging Time

If the power spectral density of a Gaussian noise signal is measured with a frequency analyser of equivalent rectangular bandwidth Δf Hz (Fig. 13) and averaged by integration over a time T, the fractional standard deviation of successive answers will be $1/\sqrt{(T\Delta f)}$. Thus if we wish to measure the spectral density at 16 Hz with a one-third octave analyser ($\Delta f \simeq 0.230 \times 16$ Hz $\simeq 4$ Hz) to a standard deviation of 5 per cent we must integrate for 100 sec according to this simple formula: if the turbulence is highly non-Gaussian (intermittent) longer times may be needed. The integration time required for a given accuracy for the mean square intensity of the total turbulence signal depends on the spectrum: remembering that the above formula gives the standard deviation in $\Phi(f)\,\Delta f$, we find that the fractional standard deviation for the density is

$$\frac{\sqrt{\left(\dfrac{1}{T}\displaystyle\int \Phi^2 \, df\right)}}{\displaystyle\int \Phi \, df}$$

where, of course, $\int \Phi \, df$ is the mean-square value itself. This formula may be useful as a rough guide if a guess can be made at the spectrum shape, but the simplest way of establishing a suitable integration time is trial and error: the scatter between successive integrations, or the excursions of a low-pass filter output, give a good indication of accuracy.

A low-pass filter with a -3 dB point at frequency $1/T$ Hz gives a standard deviation $1/\sqrt{(2T\Delta f)}$: there is little to choose between an integrator and a low-pass filter on theoretical grounds. However, my experience is that an integrator is far less tiring to use, and enables the experimenter to concentrate on the performance of the apparatus rather than on a hypnotically oscillating pointer. Commercial counting or timing circuits can be used to drive a relay to switch the integrator off after a pre-set time: times are usually chosen to have only one significant digit for convenience in calculation. It is convenient to sound a buzzer when the integrator switches off, to attract the operator's attention.

Automatic recording, by connecting a low-pass filter to a chart recorder and traversing continuously, in space or in frequency, can be used to save time at the expense of accuracy: the operator has finally to draw a smooth curve through the trace, which mixes time averaging with averaging over space or frequency, and may lead to consistent errors.

6.5. Automatic Recording of Time-average Quantities

We distinguish between digital recording of time-average values and digitization of the fluctuating signal: the latter is discussed in the next section.

The output of a low-pass averaging filter or, better, that of an integrator, can be fed to a digital voltmeter and thence recorded on paper tape, punched cards or magnetic tape either on command of the operator or at preset intervals. The difficulty comes in recording the calibration constants, like attenuator settings, integration times, filter frequencies or traverse positions, which vary from reading to reading: these must also be represented as analogue voltages and scanned by the voltmeter, or converted directly into digital form and scanned by the punch drive unit or other device, or typed in by the experimenter (with a distinct possibility of error). Since it is not possible to save much testing time by automatic recording, the length of the experiment being determined by the integration times required, the only advantage is that the data reduction can be done by computer. Unless a very large amount of work is to be done, it is probably cheaper to employ a keypunch operator to convert handwritten data rather than engage a technician to develop automatic recording apparatus. In the near future we may hope to have reading machines capable of converting arrays of numbers, or letters, written with a light pen, to corresponding arrays on magnetic tape in a computer: a cheaper alternative to this is to train the experimenter to type his results accurately and in standard format during the course of the experiment, but most people find this a considerable strain.

6.6. Digital Recording of Fluctuating Signals[45]

The complicated arrays of analogue apparatus needed for the more advanced measurements can be replaced by an analogue-to-digital converter connected to a magnetic tape-recorder or directly to a computer. If the signal can really be fed directly to a fairly large computer capable of analysing the data on line, the digital system has all the advantages of the analogue system with practically none of the worries about calibration drift, overloading and inaccuracies in the processing equipment and so on: moreover, a change from one type of measurement to another can be made by changing a few programme cards rather than by connecting, or even constructing, a new set of analogue apparatus.

In real life, however, only a few fortunate experimenters will have immediate on-line access to a sufficiently large computer. Generally, results will not be available during the course of the experiment, and at least some analogue checking of the data will be advisable to ensure reliable results: this means that digital recording is not worth while for the simpler measurements, up to, say, second-order spatial correlations, except of course for very-low-frequency signals for which analogue recording is inconvenient. The choice between an analogue multiplier costing a few hundred pounds and a digital magnetic-tape system costing as many thousands is easy to make: put another way, it is basically inefficient to record and process up to 10^5 numbers just to obtain one mean square or covariance value, and it seems unlikely that improvements in computers or even the prevalence of on-line access will make it otherwise. For more advanced measurements the cost, complication and unreliability of analogue apparatus increase rapidly and a digital system can be justified on grounds of cost-effectiveness (possibly relying on the popular fallacy that computer time does not cost real money). Therefore, we may expect digital processing to become more popular in the future as turbulence studies progress and more people attempt complicated measurements: indeed, the possibility of digital processing may encourage people to attempt measurements that would be impracticable with analogue apparatus, such as the more complicated forms of conditional probability analysis.

The sample length (integration time) required for a given accuracy is the same for digital or analogue processing; the accuracy of voltage measurement required is the same also. The total duration of the sample T is given by the results of Section 6.4: if the filter bandwidth required in a spectrum analysis is Δf and the acceptable normalized standard deviation is σ, then $T = 1/\sigma^2 \, \Delta f$; more generally, Δf can be taken as the lowest frequency for which accurate results are required, f_{min}, say.

The effective frequency response of a digital sampling system is set by the sampling rate. If the interval between samples of a sinusoidal signal of radian frequency ω is Δt, successive samples are $\sin(\omega n \, \Delta t + \phi_0)$ for $n = 0, 1, 2 \ldots$, where ϕ_0 is an arbitrary constant phase angle which we will take to be zero for simplicity. Now

$$\sin \omega n \, \Delta t = \sin(\omega n \, \Delta t + n \cdot 2\pi) \quad \text{for all } n$$

$$= \sin\left\{\left(\omega - \frac{2\pi}{\Delta t}\right)n\Delta t\right\},$$

so that a frequency ω produces the same samples as a frequency $\omega - 2\pi/\Delta t$. Remembering that positive *and* negative frequencies occur equally in random signals, we see that ambiguity can be avoided only if the numerical value of ω lies between 0 and $\pi/\Delta t$: any signals at frequencies numerically greater than $\pi/\Delta t$, say at $\pi/\Delta t + \omega'$, will be "folded" about the frequency $\pi/\Delta t$ to appear at $\pi/\Delta t - \omega'$ [strictly, at $-(\pi/\Delta t - \omega')$]. The folding frequency, $\pi/\Delta t$ or $1/(2\Delta t)$ Hz, is also called the Nyquist frequency. In brief, we must sample at more than two points per cycle of the highest frequency f_{max} in the signal (not merely the highest frequency we want to record: the signal must be passed through an analogue filter to remove any contributions above the folding frequency).

Therefore, the total number of samples required, N, is equal to the product of the sampling rate, $2f_{max}$, and the sample duration, $1/\sigma^2 f_{min}$; thus $N = (2/\sigma^2) f_{max}/f_{min}$, which depends only on the ratio of the extreme frequencies: changing the speed of the flow, or slowing down the signal by replaying an analogue tape recording at a lower speed, does not alter the number of samples required (although f_{max}/f_{min} increases with Reynolds number, being proportional to the ratio of mean flow width to Kolmogorov length scale).

Analogue-to-digital converters are digital voltmeters with a high sampling rate and a comparatively low accuracy: the latter is usually expressed in "bits", the number of binary digits used to represent a measured voltage. An N-bit A-to-D converter has a voltage uncertainty, ΔV say, of one part in 2^N of full scale. The uncertainty, known as the "quantization error", is equivalent to rounding off a decimal number, and since the rounding error is randomly distributed the output is equal to an exact digitization of the original signal plus a random "noise" signal. The probability of the apparent noise is constant in the interval $\pm\frac{1}{2}\Delta V$ and zero elsewhere so that the standard deviation of the apparent noise is $\Delta V/\sqrt{12}$; the spectral density of the apparent noise can be taken as constant up to the folding frequency and zero thereafter. Therefore, an eight-bit A-to-D converter (uncertainty 1 part in 256 of full scale) will have a root-mean-square signal-to-noise level of about 150 : 1 (remembering that full scale must be about 6 times the r.m.s. for a Gaussian signal) which is adequate for most purposes.

The output from the A-to-D converter must be now recorded on cards, paper tape or (usually) magnetic tape. The number of recording tracks on the tape must be equal to N: if N is large, the string of binary digits must be split into two or more sections and recorded in two or

more successive lines across the tape. A further difficulty arises because computers make both lateral and longitudinal parity checks on
magnetic tape: one track on the tape is filled with a binary "1" or a
binary "0" at each position to make the number of "1"s in the lateral
line odd (say), and after every few thousand lines an analogous *longitudinal* parity digit is recorded in each track to make the number of "1"s in
each track odd (say), and thereafter a sufficient "inter-record" gap
(usually ¾ in. or about 2 cm) is left for the tape to be stopped and re-
started by the computer's tape transport. Therefore (i) one track on the
tape is bespoken for parity so that a 7-track tape can accommodate only
5 binary digits and the sign, and a 9-track tape only 7 digits and the sign,
and (ii) only a few thousand samples can be recorded at once; (i) implies
that it will nearly always be necessary to split the binary number into two
parts, thus halving the effective recording speed of the tape and (ii)
implies that a data store must be used between the A-to-D converter
and the tape recorder so that the inter-record gaps can be inserted in the
tape while the A-to-D converter continues to sample at a constant rate:
the data store must have most of the attributes of a small computer and
several companies make computers in the $10,000 (£4000) price range
which are suitable for this sort of task. Large computer centres may be
able to accept continuously recorded tape without longitudinal parity
and re-record it in a format suitable for the computer. It will usually
be the tape-recorder, rather than the A-to-D converter, that limits the
sampling rate: the maximum tape speed currently available in conventional digital recorders is 150 in./sec and the maximum longitudinal
packing density 800 lines (bits) per inch, so that if two longitudinal bits
are required for each number the maximum recording rate is 60,000/sec
or a maximum folding frequency of 30 kHz, and these figures will be
reduced by up to 10 per cent by the inter-record gaps. 30 kHz is also
about the maximum frequency range of a frequency-modulation (FM)
analogue recorder running at 150 in./sec so that if signals at frequencies
higher than 30 kHz are to be recorded they must first be put on a direct-
recording tape machine and then played back at lower speed into the
A-to-D converter: direct recording—as opposed to frequency-modulation recording—is not very accurate, particularly when the speed is
changed on playback, and so 30 kHz must be accepted as the present
limit for accurate work. 30 kHz is about the highest frequency found in
turbulence measurements at subsonic speeds. If several signals, each

going up to 30 kHz, are to be recorded, then a multi-channel FM recorder can be used as an intermediary: the FM tape is replayed at a lower speed and the channels scanned in sequence by a "multiplexer" connected to the input of the A-to-D converter, so that the maximum recording rate of the final digital tape machine is not exceeded. The task of disentangling the records from each channel and of allowing for the small time difference between each channel can be performed by the data analysis program.

No general advice on data analysis programs can be given, except that care is needed to minimize computing time, particularly if the number of samples is very large. A typical computing cost is about £1 per 10,000 numbers for the calculation of a frequency spectrum: computing times can be reduced considerably by remembering that the full length of the record is *not* needed to define the higher frequencies and that the full sampling rate is not needed at the lower frequencies. For a recent review of digital processing techniques see ref. 45.

CHAPTER 7

Temperature and Concentration Measurements

Each thought he was thinking of nothing but "Snark"
And the glorious work of the day;
And each tried to pretend that he did not remark
That the other was going that way.

7.1. Separation of Velocity and Temperature Fluctuations

Even small, unintentional temperature variations can be a great nuisance when one is trying to measure velocity fluctuations: measurements of velocity fluctuations in the presence of large temperature fluctuations, such as occur in flows at high Mach number or with large mean temperature differences, require considerable care. The temperature fluctuation itself, and of course the mean temperature, can be sensed by a cold wire used as a resistance thermometer, but a *hot* wire will respond both to velocity fluctuations and to temperature fluctuations and it is necessary to separate the two, even if only the velocity fluctuation is required: in investigations of turbulent heat transfer (Section 3.9), one usually wants the mean-square velocity fluctuation, the mean-square temperature fluctuation *and* the velocity-temperature covariance. There are two main ways of separating the fluctuations.

(1) If the instantaneous signals from a "cold" wire (i.e. one carrying a current too small to heat it appreciably and operating as a resistance thermometer), say

$$e_c \equiv I_c R_c \alpha_c \theta + 0[\theta^2],$$

and a nearby hot wire, say

$$e_h \equiv au + b\theta + 0[u^2] + 0[u\theta] + 0[\theta^2],$$

155

are factored and subtracted, the output $be_c - I_cR_c\alpha_ce_h$ is proportional to u plus second-order terms: the cold wire gives θ directly so that $\overline{u^2}$, $\overline{\theta^2}$ and $\overline{u\theta}$ can be derived if the various calibration factors are known[46]. Note that changing the temperature difference between the wire and the flow changes the heat transfer from the wire (in a rather complicated fashion because k and ν both vary with temperature) so that the factor b is not simply $I_hR_h\alpha_h$ analogous to the expression for e_c; both a and b must be obtained by calibration. It is worth noting that (to first order) the difference between the outputs of two perfectly matched wires in an X probe is independent of the temperature fluctuations: thus $\overline{v^2}$ or $\overline{w^2}$ can be measured as usual, and $\overline{v\theta}$ (which is more interesting than $\overline{u\theta}$ in thin shear layers because $\varrho c_p\overline{v\theta}$ is the enthalpy flux normal to the plane of the shear layer) can be measured by using an X probe with a cold-wire resistance thermometer close by. My own feeling, based on experience, is that the use of combined hot wires and cold wires is the best technique for measurements in low-speed flows with small temperature differences: when temperature differences are large, b varies considerably and the extraction of instantaneous u by subtracting signals becomes less reliable. In high-speed flows we have the additional difficulty that the cold-wire temperature is not the fluid temperature T_f but the recovery temperature $T_f + rU^2/2c_p$ where the recovery factor r is roughly 0·9.

(2) If a single wire is operated at different temperatures[47], a and b take different values, and if these values are known the three quantities required can be obtained from three measurements of

$$\overline{e_h^2} \equiv a^2\overline{u^2} + b^2\overline{\theta^2} + 2ab\overline{u\theta} + \dots$$

The obvious course is to take one of these measurements with the wire cold so that a is zero but this is possible only in constant-current operation (in which a "cold" wire can be compensated for thermal inertia in exactly the same way as a hot wire). Constant-temperature anemometers will not run at very small mean temperature differences (unless an internally cooled probe is used) because there is no way of supplying *negative* electrical power, and the frequency response of the amplifier at small output current is likely to be poor. Care is needed to avoid spurious results due to variation of frequency response with operating temperature, whether constant-current or constant-temperature apparatus is used. The thoroughness with which the calibrations to find a and b are done depends on the ranges of mean velocity and mean temperature to be covered and upon the accuracy required—and of course on the flexibility of the test rig; it may not be possible to vary the fluid temperature at a point rapidly enough for calibration purposes, and in a supersonic wind tunnel with a fixed nozzle it is not even possible to vary the Mach number at a point. The variation of fluid properties with temperature introduces difficulties, and if the absolute temperature of the wire changes by more than 20 per cent or so the variation with temperature of the temperature coefficient of resistivity α must be taken into account. Partial or complete reliance on universal calibrations for the variation of Nusselt number with Reynolds number, temperature ratio, Mach number and so on offers an easy way out but it is even less reliable than in nominally isothermal flow because more parameters appear and the wire calibration is more likely to drift in high-speed or high-temperature flows. Most of the systematic work on measurement of combined velocity and temperature fluctuations has been done with application to high-speed flow, but wire calibration at Mach numbers more than about 1·5 is no more complicated in principle than in low-speed flow, and for the purposes of this book it is simplest to present the information so that it can be applied to either case. First, however, we examine the special problems of high-speed flows.

7.2. High-speed Flow

When appreciable fluctuations of pressure, as well as temperature and velocity, occur, the interpretation of wire signals becomes quite complicated. Also, the wire calibration becomes a function of Mach number as well as Reynolds number. The general case is discussed by Morkovin[47]: here, as an introduction to the subject and to Morkovin's report, we discuss a simplified but practical case.

The pressure fluctuations in a turbulent flow are, as remarked in Section 1.7, small compared to the absolute pressure if the fluctuating Mach number is small, which is the case at all non-hypersonic speeds. The one region in which pressure fluctuations dominate is outside the turbulence proper, where the fluctuations are entirely due to sound waves radiated from the turbulence: even here, the wire responds primarily to the resulting temperature and velocity fluctuations and the calculation or calibration of the sensitivity to sound waves is straightforward. Within the turbulence, the wire signal due to velocity and temperature fluctuations swamps that due to pressure fluctuations. Thus, the complicated discussion of fluctuation "modes", required in the case where temperature fluctuations arise both from heat transfer producing entropy changes and from adiabatic compression, is unnecessary in these two special cases.

The wire calibration can be written in a form virtually independent of Mach number for $M > 1.5$ by noting that the wire, being a bluff body, sits behind a nearly normal shock wave: thus for $M > 1.5$, the Mach number of the flow past the wire is less than 0·7 and, providing that the calibration is based on velocity and fluid properties behind a normal shock, compressibility effects are slight[48]. If this were exactly true, the temperature of a cold wire (the "recovery temperature" T_r) would be equal to the stagnation temperature, $T_0 = T_f + (U^2/2c_p)$, which is unaltered by passage through a normal shock: in fact, $T_r/T_0 \simeq 0.96$ for $M > 1.5$, $Re > 20$. The variation of Nusselt number with Mach number is negligible in this range, but the Nusselt number does depend on the wire overheat parameter $\tau \equiv (T_w - T_r)/T_r$ (as well as on the Reynolds number) because the effective fluid properties are those at some temperature between T_w and T_r. Since T_r is a nearly constant fraction of T_0, fluid properties can be evaluated at T_0 instead of T_r or the fluid temperature behind a normal shock, T_2.

In transonic flow ($0.5 < M < 1.5$) the wire calibration varies strongly with M and no comprehensive measurements have been reported: from the research point of view, interesting effects on the behaviour of turbulence do not occur below $M = 1.5$—very roughly—so the gap is not as serious as might be thought. The only way to make hot-wire measurements in transonic flow is to calibrate the actual probe to be used, over the required range of Reynolds number, Mach number and T_w/T_r: prong interference effects due to shock-wave formation rule out the use of universal calibrations.

There seems to be no practical evidence about the *highest* Mach number at which the normal-shock approximation is valid but it is at least 5. The density behind a normal shock approaches $(\gamma + 1)/(\gamma - 1)$

times the free stream density at high Mach number, so that the onset of low-density effects is postponed to a lower free-stream density than in low-speed flows: the limit is reached when the Knudsen number, (mean free path)/d, based on conditions behind the shock, rises to about 0·015, and thereafter the calibration depends on Knudsen number[49]. Of course, the Knudsen number is constant in constant-property flow. Since—according to the kinetic theory of gases—the Knudsen number is proportional to M/Re, Mach number reappears as a calibration variable in low-density flow (low-density effects have led to apparent Mach number dependence in "incompressible" flow).

7.3. Probes for Supersonic Flow

The most important requirement is reproducibility of calibration characteristics, which comes partly from a careful or fortunate choice of wire and partly from a carefully controlled process of manufacture. It is advisable for the active part of the wire to be kept well clear of the prongs because the latter carry detached conical shock waves: there will be a detached cylindrical shock over each wire sleeve (Fig. 32) so that the sleeve diameter should be kept to the minimum needed for strength and for low electrical resistance. Vibration of wires is a continuing difficulty and it is a wise precaution to measure a rough frequency spectrum with the wire in the free stream. Plug-in spectrum analysers which display the spectrum on an oscilloscope screen are useful for this purpose: this type of sweep analyser is not very suitable for low frequencies (i.e. low-speed flows). Vibration at frequencies that vary from wire to wire is attributable to excessively slack or excessively tight wires; vibration arising in the prongs or probe support will have an unvarying frequency.

At the time of writing no manufacturer unreservedly recommends his hot-wire probes for use in supersonic flow, although the more rugged hot-film probes are recommended: unfortunately there is little experience of the use of hot films in supersonic flow and, without evidence to the contrary, one might expect the frequency response to be inadequate for serious work. The frequency response required for a given minimum wavelength is, of course, directly proportional to the fluid speed: unfortunately supersonic shear layers are usually generated in small test rigs and are therefore thinner (smaller typical wavelengths) than in typical low-speed flows, so that the frequency response actually needed is further increased. There seems little hope of attaining a frequency response good enough to measure the dissipating eddies in a supersonic flow: the laser Doppler anemometer is more promising.

Cross-wire probes can be used in supersonic flows as long as the wire angle is safely less than the Mach wave angle (M > 1·4 for an angle of 45 deg) so that the wires are not affected by shock waves from the prongs. The Mach number component normal to the wire is still less than unity.

Although the advances of the last few years have nearly converted hot-wire anemometry in low-speed isothermal flows into a routine laboratory procedure, the increase in probe mortality and calibration time that must be faced in supersonic flow make it inadvisable to attempt high-speed measurements without some experience in low-speed flows.

7.4. Sensitivity of a Hot Wire to Velocity and Total-temperature Fluctuations

It is easiest to derive the sensitivity to total-temperature fluctuation, $dT_0 = dT + U \, du/c_p$: in low-speed flow $T_0 = T$, and in high-speed flow T_0 may be the more interesting quantity. The derivation below is for constant-temperature operation, assuming that the Nusselt number is independent of Mach number and that the recovery temperature ratio $\eta \equiv T_r/T_0$ is constant: the former assumption is good for Re > 2 outside the range 0·5 < M < 1·5 approximately while the latter is good for M < 0·5 (with $\eta = 1$) and fairly good for M > 1·5 (with $\eta = 0·96$) Morkovin gives the derivation for the most general case, with Nu and η functions of Re and M, and then specializes to constant-current operation.

Following Morkovin's notation[47] for the benefit of readers who wish to progress from this account to his, we have

$$I^2 R_w = \pi l k (T_w - T_r) \, \mathrm{Nu} \, (\mathrm{Re}, T_w/T_0)$$

as a definition of Nusselt number, where $\mathrm{Re} = \varrho U d/\mu$: we assume $k \alpha T^n$, $\mu \alpha T^m$ and evaluate all fluid properties at the total temperature T_0.

Differentiating logarithmically, with T_w and R_w constant,

$$\frac{d(I^2 R_w)}{I^2 R_w} = 2\frac{dI}{I} = 2\frac{dE}{E} = \frac{dk}{k} - \frac{\eta \, dT_0}{T_w - \eta T_0} + \frac{d\mathrm{Nu}}{\mathrm{Nu}}$$

with

$$\frac{dNu}{Nu} = \frac{Re}{Nu}\frac{\partial Nu}{\partial Re}\frac{dRe}{Re} - \frac{T_w/T_0}{Nu}\frac{\partial Nu}{\partial(T_w/T_0)}\frac{dT_0}{T_0}$$

and

$$\frac{dRe}{Re} = \frac{dU}{U} - \frac{dT_0}{T_0} - \frac{d\mu}{\mu}$$

(since the effect of pressure fluctuations on density fluctuations is being neglected). If I rather than R_w is kept constant a slightly more complicated expression appears. Hence

$$2\frac{dE}{E} = \frac{Re}{Nu}\frac{\partial Nu}{\partial Re}\frac{dU}{U}$$

$$+ \left[n - (m+1)\frac{Re}{Nu}\frac{\partial Nu}{\partial Re} - \frac{\eta T_0}{T_w - \eta T_0} - \frac{T_w/T_0}{Nu}\frac{\partial Nu}{\partial(T_w/T_0)}\right]\frac{dT_0}{T_0}.$$

Clearly any constant multiple of Nu can be used for calibration purposes. In particular the logarithmic derivative of Nu with respect to T_w/T_0, evaluated at constant flow conditions, can be found by measuring the wire current or voltage at a number of values of resistance, if the variation of wire resistance with temperature is known: we have

$$I^2 = \pi lk(T_w - \eta T_0)\,Nu/R_w$$

with

$$R_w = R_1(1 + \alpha_1(T_w - T_1) + \beta_1(T_w - T_1)^2)$$

where suffix 1 indicates a reference temperature, say 0°C. Morkovin recommends

	α_1	β_1
Tungsten	$4\cdot0 \times 10^{-3}$	$8\cdot0 \times 10^{-7}$
Platinum	$3\cdot8 \times 10^{-3}$	$-6\cdot5 \times 10^{-7}$
90/10 platinum/rhodium	$1\cdot6 \times 10^{-3}$	$-1\cdot54 \times 10^{-7}$.

From the above equation for I^2

$$\frac{2}{I}\frac{\partial I}{\partial T_w} = \frac{1}{T_w - \eta T_0} + \frac{1}{Nu}\frac{\partial Nu}{\partial T_w} - \frac{1}{R_w}\frac{dR_w}{dT_w} = \frac{2}{I}\frac{\partial I}{\partial R_w}\frac{dR_w}{dT_w}$$

where $dR_w/dT_w = R_1(\alpha_1 + 2\beta_1(T_w - T_1))$ and $\partial I/\partial R_w$ is to be evaluated at constant flow conditions. Finally, then

$$\frac{T_w/T_0}{Nu} \frac{\partial Nu}{\partial(T_w/T_0)} = \frac{T_w}{R_w} \frac{dR_w}{dT_w} \left(\frac{2R_w}{I} \frac{\partial I}{\partial R_w} + 1 \right) - \frac{1}{1 - \eta T_0/T_w}$$

for substitution in the above sensitivity equation. It is advisable to obtain α_1 and β_1 from measurements on a sample of annealed wire from the same reel as that used for the test probes: properties of fine wires may be different from that of the bulk material.

A slightly easier conversion from resistance to temperature appears if we fit the resistance–temperature data to a power law, say $R_w/R_1 = (T_w/T_1)^p$. Then

$$\frac{T_w/T_0}{Nu} \frac{\partial Nu}{\partial(T_w/T_0)} = p \left(\frac{2}{R_w} \frac{\partial I}{\partial R_w} + 1 \right) - \frac{1}{1 - (R_r/R_w)^{1/p}}$$

where R_r is the recovery resistance, i.e. the resistance of the unheated wire. $p \simeq 1\cdot 1$ for tungsten, $0\cdot 98$ for platinum, and $0\cdot 67$ for a nickel hot film in the range 0–300 °C. Unfortunately platinum–rhodium and other alloys deviate so far from the pure-metal appoximation $R \propto T$ that a power law is not a good fit over a 2 : 1 range in absolute temperature.

The calibration procedure for measurements in flows with large temperature fluctuations therefore involves measurement of Nu as a function of Re at one overheat ratio and measurement of I as a function of R at several Reynolds numbers: measurement of I as a function of R at constant Re can be done very quickly if the flow is steady, whereas changing the Reynolds number takes longer. In a supersonic tunnel, Re can be changed only by changing the total pressure, which may well change the total temperature as well so that Nu as a function of Re at given overheat ratio may have to be obtained by interpolation. In other cases also, one may have to accept lower standards of accuracy than in isothermal low-speed flows. Naturally, wire breakage is more common in high-speed flows at high stagnation pressure, which makes prolonged calibration procedures unrealistic. If a universal calibration for Nu as a function of Re is to be relied on, heat loss to the supports must be allowed for: Morkovin discusses this, and a useful summary of the whole operating procedure, including an allowance for the variation of η with Re, is given by Vrebalovich[50].

The only fundamental assumption made in the above analysis is that the Nusselt number, as defined by the first equation in this section, is a function of Re and T_w/T_0 only. The neglect of M has been discussed: the neglect of T_w and T_0 as separate parameters of Nu is justifiable if the power-law assumptions for k and μ are accurate (it is, of course, the use of power-law formulae for k and μ that enables us to determine the temperature sensitivity by altering the temperature of the wire rather than that of the flow). Morkovin recommends $n = 0.885$, $m = 0.765$ for the range 270–350 °K, but other data (e.g. Schlichting[11], table 14.1) suggest that n is much nearer m, not exceeding 0.80, which agrees better with the received doctrine that c_p and $\mathrm{Pr} \equiv \mu c_p/k$ are nearly independent of temperature.

Morkovin and other workers have used an elegant "fluctuation diagram" technique for fitting the best curve of the right theoretical shape to the experimental points. Since this sort of data reduction is nowadays done on a computer, a less elegant but more straightforward procedure suffices. Writing the calibration equation for the case of negligible pressure fluctuations as

$$\frac{dE}{E} = A\left(X\frac{dU}{U} + \frac{dT_0}{T_0}\right)$$

we have

$$\frac{\overline{(dE)^2}}{A^2E^2} = X^2\frac{\overline{(dU)^2}}{U^2} + 2X\frac{\overline{dU\,dT_0}}{UT_0} + \frac{\overline{(dT_0)^2}}{T_0^2},$$

so that if we plot the left-hand side against the sensitivity ratio X at a number of wire temperatures we define a parabola: the three turbulence quantities can be deduced from a least-squares or other type of fit to this parabola. As usual when using curve-fitting procedures on a computer, the standard deviation of the points (or the answers obtained after rejecting the worst point) should also be printed out.

7.5. Small Temperature Differences

The procedure outlined above is intended for use in flows with large temperature differences: formally, it is still the right procedure to use when correcting hot-wire sensitivity for changes in test-rig temperature

(Section 5.2). Calibrations of I against R_w at a few Reynolds numbers can be used to deduce $\partial \mathrm{Nu}/\partial T_0$ and $\partial^2 \mathrm{Nu}/\partial \mathrm{Re}\, \partial T_0$ so that

$$\frac{\partial}{\partial T_0}\left(\frac{\mathrm{Re}}{\mathrm{Nu}}\frac{\partial \mathrm{Nu}}{\partial \mathrm{Re}}\right)$$

follows. In practice, rudimentary calibrations on wires similar to the one in use will suffice for corrections in low-speed test rigs. If temperature changes during a run are large enough for more careful calibration to be required, temperature *fluctuations* may be large enough to affect the readings. Arya and Plate[51] used a simplified calibration procedure for quite large temperature differences.

In liquids, viscosity varies very rapidly with temperature and, although the thermal conductivity varies much less rapidly than the viscosity, the variation is not negligible: for water, $m \simeq -5$ and $n \simeq 0.7$. Temperature differences are usually smaller than in gases so that β_1 can often be neglected. Also, $\varrho \simeq$ constant so $m + 1$ is replaced by m in the sensitivity formula. Apart from this the procedure is exactly the same as in gases: the only difficulty that does not appear in gases is that the wire temperature is severely limited by bubble formation so that the range of T_w/T_0 available for calibration is small and the required accuracy of resistance measurement correspondingly greater.

7.6. Measurements in the Presence of Concentration Differences

Concentration differences can be dealt with in the same way as temperature differences[52]. A simplification is that Nu at given Re ought not to depend on concentration, unless the indices m and n vary greatly with concentration, because the concentration, unlike the temperature, is not altered by the presence of the wire. Nu will, of course, depend indirectly on concentration because of the change of viscosity appearing in the Reynolds number.

For constant-temperature wire operation in an isothermal flow of concentration c, we have, by an analysis similar to that used for temperature,

$$2\frac{\mathrm{d}E}{E} = \frac{1}{k}\frac{\mathrm{d}k}{\mathrm{d}c}\,\mathrm{d}c + \frac{\mathrm{Re}}{\mathrm{Nu}}\frac{\partial \mathrm{Nu}}{\partial \mathrm{Re}}\left(\frac{\mathrm{d}U}{U} + \frac{1}{\varrho/\mu}\frac{\mathrm{d}\varrho/\mu}{\mathrm{d}c}\,\mathrm{d}c\right).$$

If $(\mathrm{Re}/\mathrm{Nu})(\partial \mathrm{Nu}/\partial \mathrm{Re})$ varies sufficiently with wire temperature we can make measurements at three wire temperatures and deduce $\overline{(dc)^2}$, $\overline{(dU)^2}$ and $\overline{dc\,dU}$. An alternative procedure is to make simultaneous measurements with two adjacent wires of different diameters[53], having different Reynolds numbers and different values of $(\mathrm{Re}/\mathrm{Nu})(\partial \mathrm{Nu}/\partial \mathrm{Re})$ (note that the wire diameter cancels out of this expression), but the resulting equations for U and c are not very well conditioned. Way and Libby[54] have recently developed a cunning technique in which a wire or film is mounted in the wake of another, both being operated at constant temperature. The heat transfer from the hot wake to the downstream wire varies more strongly with the thermal conductivity of the fluid than the heat transfer from the upstream wire, which responds primarily to velocity, as usual. Therefore the velocity and thermal conductivity (and hence the concentration) can be extracted by solving two well-conditioned equations containing the two wire voltages. In gases, the variation of μ and k with concentration can be derived from semi-empirical arguments based on kinetic theory. The more common liquid mixtures or solutions (e.g. salt and water) are well documented empirically.

Measurements in the presence of concentration *and* temperature differences can, in principle, be made by running at six different wire temperatures, or by running two or more wires simultaneously at a smaller number of successive wire temperatures.

An alternative hot-wire instrument, for measuring concentration in an isothermal flow directly, is the katharometer, in which a known rate of mass flow of the mixture is drawn through a tube containing a hot wire, whose heat loss is a function only of the concentration. For gases, a technique which holds out the possibility of a good frequency response is to extract the sample through a probe with a small sonic nozzle well upstream of the wire: the mass flow rate is not constant but the heat loss is still a function only of concentration. The gas must be fairly dry to avoid condensation in the sonic nozzle.

Care may be needed to avoid chemical or electrolytic reaction between the wire and the mixture. Salt water is notoriously corrosive, for instance, and platinum catalyses some hydrocarbon–oxygen reactions. A particularly annoying phenomenon has recently been found in air–helium mixtures (Tombach[25]). There is a large and erratic temperature jump at the surface of a heated metal exposed to helium; also, a wire exposed to a high concentration of helium and then returned to a low or zero concentration of helium takes several hours to outgas the helium adsorbed onto the wire surface; it appears that this effect may not be too serious if helium is used only in low concentrations (i.e. as a tracer).

Concentration in liquid mixtures or solutions can be measured by an electrical conductivity probe[55]: the current flowing between a very small electrode and a larger one some distance away depends almost entirely on the electrical conductivity of the liquid near the small electrode (typically the end of a platinum wire imbedded in a glass probe). Spatial resolution and frequency response can be as good as that of a hot-wire anemometer. Calibration is needed, and it is important to keep the electrode clean.

Dye concentration in liquids or smoke concentration in gases can be measured by attenuation of a light beam[56]. By using fibre-optics techniques, good spatial resolution can be obtained. It is difficult to produce smoke of consistent quality, so that fluctuation measurements are best referred to the local mean concentration measured at nearly the same time: dyes are easier to produce in repeatable strengths.

Light *scattered* by smoke or dye also provides a measure of concentration[57]; the optical arrangement is similar to that of the Doppler anemometer but of course the amplitude, rather than the frequency, of the photocell output is measured. The advantage over the light-attenuation method is that good spatial resolution can be obtained without minute optical components: the advantage of having no solid body in the flow is balanced by the need to locate the point at which the detector is focused onto the light beam.

Readers are referred to the original research papers for fuller discussions of these techniques. Concentration measurements may be useful in basic research on turbulence: for instance, Fiedler and Head[58] measured the intermittency of a turbulent flow by detecting the concentration boundary rather than the vorticity boundary.

CHAPTER 8

Summary of Practical Details

"But oh, beamish nephew, beware of the day,
If your Snark be a Boojum! For then
You will softly and suddenly vanish away,
And never be met with again!"

THIS chapter is partly a recapitulation of the material given in the earlier chapters on measurement techniques and partly a collection of miscellaneous advice. It is intended primarily for the man who will operate the apparatus, although the man who will make use of the results may find it helpful in estimating their accuracy. Sections 8.1 to 8.3 refer specifically to the hot-wire anemometer: the rest of the chapter is more general.

8.1. Choice of Anemometer (Section 4.2)

If the r.m.s. turbulence intensity lies between 0·1 and 10 per cent of the local mean velocity, a range which covers nearly all turbulent flows with the exception of wind-tunnel free-stream turbulence and jets or separated flows, then either a constant-temperature or a constant-current hot-wire anemometer can be used. If the turbulence intensity is less than 0·1 per cent then a constant-current anemometer, with an amplifier having very low noise or a transformer-coupled input, is advisable and if the turbulence intensity exceeds about 10 per cent, then a linearized anemometer should be used: in practice this means a constant-temperature anemometer. If quantities relating to the skewness of the probability distribution, i.e. triple products, are to be measured then the anemometer should always be linearized. For turbulence levels less than about

167

10 per cent, the use of a constant-current anemometer with a voltage linearizer only would be acceptable because the effect of fluctuation in time-constant would be fairly small: however, this arrangement is seldom used in practice. The exact turbulence level at which a linearizer becomes essential depends upon the quantities being measured; for measurements of mean square fluctuation intensity only, it is doubtful whether a linearizer really adds accuracy to the result, because it is, of course, an extra complication and an extra source of error in itself. It is certainly advisable to use a linearizer if spectra or correlations are to be measured in high-intensity turbulence, but it must be remembered that large changes in flow direction will introduce inaccuracies, due to prong interference and uncertainty of wire response, which cannot be compensated by linearizing the U-component calibration.

8.2. Choice of Probe (Sections 5.3, 5.5)

For measurements of lateral fluctuations a compromise must be reached between the use of a large number of wires and the need for a large number of readings. For instance, $\overline{uv}$ could be measured with either one, two or three wires: three wires could be arranged in a "cross-wire" probe, with a normal wire by the side, to measure u and v separately; two wires could be arranged in a cross and u and v extracted by means of summing and differencing the two signals; and a single wire could be set at alternately positive and negative angles to the flow so as to measure $\overline{(u + v)^2}$ and $\overline{(u - v)^2}$, whose difference is proportional to $\overline{uv}$: finally, a single wire could be (slowly) rotated[59]. The use of three wires avoids the inaccuracy of deducing the u component fluctuation from inclined wires; two wires permit the required quantity to be obtained as a single reading; and the use of one wire avoids the difficulty of matching two wires in an X probe, but a mechanical device must be used to rotate the probe and the final product $\overline{uv}$ must be obtained as the difference of two readings taken at slightly different times. It is, of course, always undesirable to obtain the final answer by subtracting hot-wire readings taken at long intervals, because of calibration drift: however, calibration drift due to temperature changes or dirt deposition will tend to affect all the wires in the same sense so that the percentage drift in the difference between two wire signals is generally no worse than the percentage drift of either wire signal separately. Answers obtained from a slightly mismatched X probe can be improved by taking the average of the original reading and one obtained after rotating the probe 180° to interchange the wires.

It is possible to measure almost any quantity with a single wire by inclining it at a sufficiently large number of different angles to the flow but this is not usually an accurate procedure for the more complicated measurements. Some workers with only a single anemometer have measured the v component by finding the sum of $\overline{(u + v)^2}$ and $\overline{(u - v)^2}$ with a wire set at $\pm 45°$ to the flow and then subtracting the value of $\overline{u^2}$ obtained with the wire set normal to the flow: this technique may sometimes be used when the size of the probe must be kept as small as possible. Occasionally[38] the v component has been measured by using a "cross-wire" probe with one wire at 45° to the air stream and the other *normal* to the air stream, the advantage being that u can be measured with more assurance by the normal wire than by taking the sum of the signals of two inclined wires. This technique is particularly suitable for measuring $\overline{uv}$, but unfortunately commercial probes with this configuration are not available.

General practice is to use a cross-wire probe ("X" probe) with the wires set at roughly $\pm 45°$ for v-component measurement, and an "X" probe or a single inclined rotating probe for $\overline{uv}$. Triple products can be measured with these simple types of probes: $\overline{v^3}$ can obviously be measured by taking the cube of the difference between the signals from two wires of an "X" probe and once $\overline{v^3}$ is known $\overline{u^2v}$ can be deduced from cubes of the signal from a single inclined wire set at alternately positive and negative angles to the flow, i.e. $\overline{(u \pm v)^3}$. $\overline{w^2v}$ can be deduced from the cube of the difference between the two signals from the wires of an "X" probe, with the plane of the wires bisecting successively the positive and negative angles between the y- and the z-axes, giving $\overline{((v \pm w)/\sqrt{2})^3}$.

The angle between the probe axis and the direction of the mean velocity should be kept the same in measurement runs as in calibration. According to cosine law, an "X" probe will indicate fluctuating velocities resolved with respect to its axis rather than the stream direction if the two are different, but it is unwise to extrapolate the yaw calibration of the probe—that is, if a cross-wire probe is calibrated within a yaw angle range of $\pm 5°$, it should be kept within $\pm 5°$ of the stream direction and the appropriate yaw sensitivity used in reducing the results. If the probe is not linearized, its sensitivity will alter as the mean velocity component normal to the wire alters, which is another reason for keeping the axis close to the stream direction. Of course, any axis can be chosen for calibration (it is sometimes useful to yaw an X probe slightly in order to match the wires) but angles of more than 5° or so between the stream direction and the centre line of the probe body may cause appreciable prong or body interference with the flow round the wires. Several authors have remarked on large differences in the calibration of a U probe according to whether it is inserted with the axis along or perpendicular to the direction of the mean velocity. Workers whose only concern is with thin shear layers tend to be rather careless about interference between a probe and its support simply because this type of interference generally has very little effect on a Pitot tube, because the total pressure on a given streamline (unlike the velocity) is not altered by interference. A useful question to ask oneself is whether one would believe the readings of a *static* tube or small anemometer placed at the position of the wire. This is also a useful question to ask when designing the probe or when choosing a commercially made probe. Wires for correlation measurements should be matched (same length and diameter) because although differences in mean calibration and thermal-inertia time constant are allowed for by the usual operating procedures, the heat conduction to the supports produces amplitude and phase errors whose effects on correlations are minimized by making the wires identical. Also, the spatial resolution depends on wire length and should be made the same for both wires.

Practical details of film design are discussed by Bellhouse and Bellhouse[60].

8.3. Calibration (Section 5.1.3)

Wherever possible probes should be calibrated in the test rig in which they are to be used, to avoid errors due to fluid temperature differences. If the velocity is not uniform across the test rig, it may be necessary to use a substitution technique, first measuring the velocity at the proposed calibration station with respect to some test-rig reference pressure, and then calibrating the probe with respect to the test-rig pressure. Apart from the difficulty of taking mean readings, there is no reason why a hot-wire probe should not be calibrated in a turbulent stream of any intensity up to that at which serious doubt arises about the accuracy of velocity measurements by Pitot tubes. The anemometer voltage can conveniently be plotted against the velocity or flow angle in a form chosen to give a straight line (5.1.2, 5.3.2). Some correction for fluid properties will have to be made when reducing the results: even if the wire calibration is not to be plotted as Nusselt number versus Reynolds number, it will still be necessary to calculate the density of the flow in order to deduce the velocity from Pitot tube measurements. If large fluid temperature changes (more than about 5°C) are expected, it is advisable to do calibrations at different temperatures to establish or check the temperature dependence of the calibration. The number of points needed on a calibration depends, of course, on the accuracy of measurement but since one of the symptoms of a malfunctioning wire is a slight departure from a straight line on the King's law or yaw response plot, it is advisable to measure at least six calibration points on a new wire to make sure that its calibration really is straight. If the air is dirty or the temperature variable, frequent calibration checks may be needed; the only guide here is experience, but it must be remembered that, even in the hands of an expert, hot wires are subject to so many outside influences and the results are so difficult to cross check that careful calibration checks are *essential*. A good way of checking a new hot-wire system is to measure the Reynolds shear stress in a pipe (where the radial shear stress gradient is equal to the longitudinal pressure gradient). Any other well-documented measurement like the u-component r.m.s. turbulence intensity at the centreline of a pipe, known to be about 0·8 times the friction velocity u_τ, can be measured instead. If one is confident in the general hot-wire system, one can use this sort of technique for calibrating individual wires, but it is normally best to do the initial calibration explicitly and use repetitions of known turbulence measurements only as an occasional check on the overall accuracy.

It is extremely difficult to check whether the slope of an allegedly straight line is really constant over the whole length of the line. The example in Fig. 45 shows a set of points which lie on a parabola whose slope is as much as +10 per cent different from the slope of the straight line even though the maximum deviation in the ordinate is only just over 1 per cent of full scale. Figure 46 shows a similar example: the points represent a plot, against $U^{0.5}$, of a hot-wire calibration that

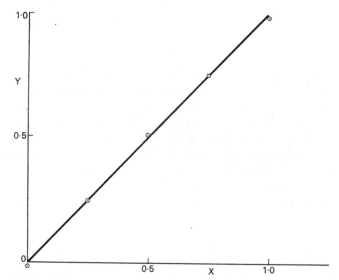

FIG. 45. The line is $Y = X$. The points lie on $Y = X + 0 \cdot 0125 - 0 \cdot 1(X - 0 \cdot 5)^2$, which deviates only $\pm 0 \cdot 0125$ in ordinate but $\pm 0 \cdot 1$ in slope.

really varied as $U^{0 \cdot 45}$. There is an appreciable difference between the slope of the curve through the points and the slope of the best straight line, even though the scatter of the ordinates about the best straight line is very small. Admittedly, the speed range between $U^{0 \cdot 5} = 0 \cdot 25$ and $U^{0 \cdot 5} = 1$ is a ratio of $16 : 1$ which is rather larger than is expected in most experiments: for smaller speed ranges, the errors would be smaller. However, Fig. 45 could refer to any part of a linearizer calibration curve, and its implication is that the slope of a linearizer calibration cannot be measured to better than ± 10 per cent except by taking extreme care. This means that great reliance must be placed on the accuracy of the linearizer when making intensity measurements over a wide range of velocities, which implies that it is not very sensible to use a linearizer merely to make the arithmetic of reducing the results easier. The proper purpose of a linearizer is to reduce the curvature of the wire response, and here any improvement on the original wire calibration is better than none.

If the hot-wire calibration is $E^2 = A + BU^{0 \cdot 45}$, the required transfer function of the linearizer is $E_{\text{out}} = c(E_{\text{in}}^2 - A)^{2 \cdot 22}$. Usually the function

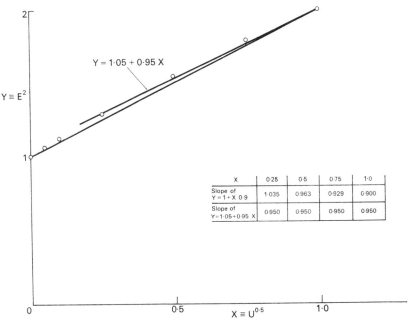

X	0.25	0.5	0.75	1.0
Slope of $Y = 1 + X^{0.9}$	1.035	0.963	0.929	0.900
Slope of $Y = 1.05 + 0.95 X$	0.950	0.950	0.950	0.950

FIG. 46. The points lie on $Y = 1 + X^{0.9}$ (i.e. $E^2 = A + BU^{0.45}$). The line $Y = 1.05 + 0.95X$ is a best-straight-line fit for $0.25 < X < 1$ (a speed range of 16 : 1).

generator itself will be set for only one value of A, A_0, say, so that the anemometer voltage E must be amplified or attenuated by the factor $\sqrt{(A_0/A)}$ to match it to the linearizer. Therefore A must be found by calibration: as mentioned in Section 5.1.3, it is unwise to find A by measuring the heat transfer in still air because the "King's law type of calibration" quoted above is not valid down to $U = 0$. It is therefore advisable to find A from a full calibration of the wire just as if no linearizer were being used: this takes some strength of mind, but the alternative is to accept that the linearizer output will be slightly non-linear by an amount which (see above) is rather difficult to measure directly. The output slope of a correctly adjusted linearizer can be calculated as $c(A_0 B/A)^{2.22}$ or, more safely, found from a few experimental points.

If a linearizer is not being used and if only correlations, spectra or other measurements of dimensionless quantities are being made, complete calibration of the wires is not necessary. For instance, in measurements of spectra and correlations of $\overline{u^2}$ and $\overline{uv}$, no calibration at all is needed. For v-component spectrum or correlation measurements, it is only necessary to match the u-component sensitivities of the two wires of the X-probe so that subtracting the two signals will give a voltage proportional to v; that is, the mean voltage difference between the two wires should be independent of speed over a small range near that which the measurements are to be made. The most general way of doing this is to linearize both wires and attenuate the output of the linearizer having the larger slope: it is not possible to match the calibrations of unlinearized wires over a range of speeds unless the wires are identical, but a "local" match can be achieved by varying the test-rig speed slightly above and below the operating speed and attenuating one wire output until the difference is independent of speed. The yaw calibration of a wire, recorded as the tangent of the effective angle, should not vary during the life of the wire unless the wire bends (poor joints, a loose wire or thermal expansion) or gets extremely dirty. However, the difficulty of measuring the slope of a yaw calibration is considerable (Fig. 45 applies once more) and if good results are required it is a good idea to make at least two determinations of the effective angle, which can conveniently be done at the beginning and at the end of the set of measurements.

One of the advantages of the hot-wire anemometer is that it can give accurate mean-flow measurements (in isothermal flows) at speeds below those at which a Pitot tube can be relied upon. It is permissible to make a moderate downward extrapolation of the calibration of a hot wire against a Pitot tube, but for speeds less than about 30 cm s^{-1} in air it is undesirable to use either an extrapolated calibration or a "universal" calibration. The simplest method of calibrating a wire at such low speeds is to put it into the settling chamber of a wind tunnel or similar test rig and measure the speed by means of a Pitot tube in the working section. A rough allowance will generally be necessary for the displacement thickness of the boundary layers on the walls of the working section, and care should be taken to check that the flow in the settling chamber is sufficiently uniform, but a speed reduction of 10 : 1 can easily be obtained by this means. For very low speeds it may be possible to move the probe with respect to stationary fluid, either on a straight track or whirling arm, or by vibrating the probe. The sensitivity to velocity fluctuations can be found without measuring the mean speed by subjecting the hot wire to a sound field of known intensity. The root mean square particle velocity is proportional to the root mean square sound pressure level, the relation between the pressure fluctuation p and the velocity fluctuation u being given by $p = \varrho au$ where a is the speed of sound. For moderate sound fields, the hot-wire response is due primarily to the velocity fluctuations with no appreciable contribution from the temperature fluctuations in the sound wave. In liquids, the mass flow in the calibration rig can be found by direct weighing: a direct displacement technique can be used in air also, displacing air from a closed container through a nozzle by admitting water at a constant rate through a pipe. Perhaps the most basic method of all is to move a container full of liquid bodily past the probe[61].

The special problems of hot wire or hot-film calibration in liquid mercury, where the probe is almost inevitably contaminated each time it passes through the surface, are discussed by Malcolm[62] and Hoff[63].

8.4. Errors

8.4.1. *Spurious Signals*

The causes of spurious signals are:

(1) *Eddy shedding from the probe.* This is usually confined to a narrow band of frequencies and is very dependent on the angle between the probe axis and the direction of the mean velocity, which provides both a means of diagnosis and, generally, a means of cure. Providing that the front of a hot-wire probe is well streamlined, only the prongs are likely to shed eddies strong enough to affect the wire: eddy shedding from the rear of the probe body or its support is not usually important.

(2) *Vibration or "strain gauging".* The natural frequency of the probe support or a hot wire itself is almost independent of flow speed although the amplitude may depend considerably on speed: the remedy is to make the probe support sufficiently robust and to avoid tightly tensioned wires (the greater the amount of slack in the wire the smaller the change in wire tension required to provide a given acceleration). "Strain gauging" is not a problem with hot films, but it is frequently encountered

when using hot wires at high speeds. Vibration caused by the test-rig drive (for instance, vibration at the fan rotation frequency or blade passage frequency) can affect all types of anemometer: the laser Doppler anemometer, with its long optical path lengths, is particularly vulnerable.

(3) *Electronic oscillations.* Oscillations generated within the electronic apparatus itself are usually of very high frequency and therefore fairly easy to remove by inserting suitable capacitors. Commercial constant-temperature anemometers usually have variable inductances or capacitors which can be adjusted to keep the feedback loop stable. Occasionally, interference from long-wave broadcasting stations is a nuisance but hot wires and films are very inefficient antennae because of their low resistance. Electromagnetic radiation from nearby transformers and rectifiers is more troublesome but easier to control. Very puzzling effects can be produced if there is appreciable radiation from the test-rig drive, but this can be identified by running the test-rig with the probe covered. Sometimes what appears to be a random-noise signal when viewed on a slow-speed oscilloscope trace may prove to be a narrow-band high-frequency oscillation. Small high-frequency oscillations can be filtered from the output but larger oscillations should be cured at source, as must any oscillation within the frequency range of the turbulence. Mains frequency "hum" is the most important of the latter. Any hum that appears when connecting together two pieces of apparatus whose separate performance if satisfactory is probably due to "earth loops", that is, excessive resistance in the earth connections.

(4) *Electronic noise.* The only cure for broad-band noise is proper circuit design, although a gradual increase in the noise from a given piece of apparatus indicates that some component is dying. To a good approximation, electronic noise is uncorrelated with the turbulence signal so that its mean square (not root mean square!) can be subtracted from the total mean square signal to give the mean square of the turbulence signal alone. In spectrum measurements the noise in each frequency band must be subtracted: usually the noise is appreciable only in the higher frequencies where the turbulence signal is falling off rapidly. An elegant method of eliminating large noise signals has been suggested by Poreh et al.[64]: if two hot wires are placed very close together their turbulence signals will be perfectly correlated whereas the

noise from the two chains of electronics will be uncorrelated; therefore the mean product of the anemometer outputs will be the mean square of the turbulence signal. Poreh *et al.* also deal with the more practical case where the wire spacing is finite and the correlation imperfect: their technique is particularly useful for measuring space or time derivatives of fluctuating quantities. Noise is particularly troublesome when the signal has to be differentiated, thus accentuating the high-frequency contribution. In all cases, noise can be minimized by inserting a low-pass filter with a cut-off frequency just above the expected upper limit of the turbulence spectrum.

(5) *Test-rig turbulence.* The background turbulence in wind tunnels, and other test-rigs intended to produce a uniform stream, is usually small enough for its mean square to be subtracted from the output, although low-frequency undulations and unsteadiness can cause spuriously large spreading rates in wakes by moving the whole shear layer from side to side. If measurements are being made near the wall of the test-rig it should be remembered that the background turbulence level is usually larger there, due to the remains of the test-rig boundary layer. It is not usually possible to distinguish between test-rig turbulence and true turbulence in duct flows, unless the flow is nominally uniform at the entrance to the measurement section.

(6) *Temperature fluctuations.* If the mean temperature of the test-rig changes appreciably with time, or if such changes are prevented by a heat exchanger a short distance upstream of the measurement station, temperature fluctuations may occur. For many purposes, their effect on a hot-wire or hot-film probe may be treated as part of the test-rig turbulence, but again the intensity is likely to be larger near the walls than in the central part of the flow, due to heat transfer through the test-rig walls. The r.m.s. temperature signal is directly proportional to the mean voltage across the probe and can be measured quite easily by putting a small current through the probe so that it responds only to temperature and not to velocity (the sensitivity to velocity fluctuations varies as a high power of probe current).

The formal way of discovering whether spurious signals exist is to measure a frequency spectrum, preferably in the free stream of the test-rig: in practice, observation of the signal on an oscilloscope is frequently sufficient.

Distortion of the signal by the probe itself is difficult to assess, even if prong interference is minimized and a linearizer used, because correction of a low-order mean product for distortion requires knowledge of higher-order mean products (5.3.1). The most recent discussions of the problem are by Guitton[65] and Rose[66]: one has to accept that the precentage accuracy of measurements with probes becomes worse as the turbulence level increases, and in particular that reversals of flow direction are not sensed by hot wires, which "rectify" the signal. Hot films are probably less sensitive in reversed flow, and Pitot tubes in reversed flow indicate approximately the static pressure.

8.4.2. Drift

(1) *Drift in the electronics.* This can be checked before the experiment: usually the only quantity that needs to be monitored during the experiment is the zero error of any d.c. coupled apparatus. Reduction of drift in a given piece of apparatus is, of course, a matter of circuit design although drift can be minimized by keeping the temperature of the apparatus constant—for instance, by fitting it in a temperature-controlled cabinet.

(2) *Drift in the probe calibration.* This is the curse of hot-wire and hot-film anemometry. Drift due to physical or chemical changes in the probe can be minimized by careful manufacture and by running new probes at their maximum temperature and the maximum speed of the test-rig until the cold resistance (referred to a given temperature) has ceased to change. Immersion of the probe in an ultrasonic cleaning bath is a useful preliminary test. If the probe calibration appears to be changing during use, the first thing to do is to check the cold resistance at the fluid temperature: if it has changed, while the resistance referred to a standard temperature has not changed, then the calibration drift is probably due to fluid temperature changes. If the cold resistance has not changed, then any change in the calibration is probably attributable to dirt on the probe, although conventional causes like leaks in the manometer circuit or changes in the test-rig circuit should not be ruled out. A dirty probe should be cleaned, either by a strong organic solvent or, preferably, an ultrasonic cleaning bath, and then recalibrated: hot-film probes—or even wires—can be cleaned with a soft brush but the cold resistance should be checked afterwards to make sure that no

mechanical damage has occurred. If the new calibration differs from the first calibration by more than a few per cent the probe should be discarded because appreciable dirt deposition changes the frequency response of the probe as well as the mean calibration. The simplest way of checking the calibration is to repeat a reading, either the first reading of the run or a standard turbulence reading at some reference position in the test-rig.

Both Richardson and McQuivey[67] and Resch and Coantic[68] have found that a dirty hot film in water will have the same calibration as a clean one at a lower current or temperature ratio: this reduces recalibration to the one point necessary to identify the "clean" calibration. This is an empirical rule without any physical justification and, like all such rules, it should be used with caution. Moreover, it should not be used to compensate for large calibration changes because of the effect of dirt on the frequency response.

Solutions of long-chain polymers (Section 3.10) have some resemblance to dirty fluids and some very curious results have been obtained both with heated-element anemometers and with Pitot tubes. Calibrations by Astarita and Nicodemo[69] have shown that results are repeatable but that the sensitivity of hot film probes in much smaller than in the pure solvent. This finding is not necessarily valid for *all* non-Newtonian fluids of this sort.

8.4.3. Inaccuracy of the Electronics

Apart from changes in calibration due to changes in ambient temperature or mains voltage, the main source of difficulty with electronic apparatus is overloading (distortion) which is easy to overlook if one has a long chain of processing equipment. The maximum permissible peak-to-peak input should be marked on each piece of apparatus and checked periodically during the course of the experiment. Note that it is the peak rather than the r.m.s. signal that matters: the peak-to-peak value of a Gaussian signal is about 6 times the r.m.s. but this figure may be considerably exceeded by intermittent signals, and highly skewed signals have large peak values on one side only.

It does not pay to be too conservative in setting input levels because too small a signal level degrades the signal-to-noise ratio: also, many non-linear devices such as biased-diode function generators perform rather poorly at low input levels. The need to maintain a high but not excessive input level to each piece of apparatus means that attentuators must be fitted at several points in the chain of apparatus. Where possible these should be based on pre-calibrated fixed resistors, brought into circuit by a multi-position switch, rather than by infinitely variable potentiometers. Even if the latter are individually calibrated, the wiper contact resistance (or its position!) may vary with time. If infinitely variable potentiometers must be used it is a good idea occasionally to move them rapidly through their full-scale range of travel several times in order to clean the wiper contacts. If several attenuator settings have to be multiplied together when working out the results, it is simplest to calibrate them in decibels or in powers of $\sqrt{2}$. Calibrated attenuators should normally be provided with their own input and output amplifiers (emitter followers or cathode followers) in order to give acceptable input and output impedances. Referring to Fig. 42, it can be seen that in order for the effective gain of the attenuator to be equal to the gain as measured across the input and output terminals, it is necessary for the source resistance to be much smaller than R_1 or R_2 and for the load resistance to be much larger than R_1 or R_2. Of course, failure to satisfy these inequalities does not matter if the attenuator is not to be calibrated. It is sometimes convenient to insert simple low-pass or high-pass filters, consisting (Fig. 42) of one resistor and one capacitor, in the circuit. These may also produce an attenuation at medium frequencies if the above inequalities are not satisfied. Wherever possible, the whole chain of apparatus should be calibrated together by inserting a known input signal and measuring the output signal. It is usually sufficient to measure the peak-to-peak level of a sinusoidal input on an oscilloscope providing that the oscilloscope calibration is occasionally checked with reference to a d.c. voltmeter. Attempts to calibrate the whole chain of apparatus to better than 1 per cent are not really justified in view of the unreliability of the basic sensing elements, but the overall error of a long chain of apparatus can be alarmingly larger than the error in individual units.

Wherever possible, exact calibration of the wire and of the electronics should be avoided by measuring dimensionless quantities like correlation coefficients or skewness coefficients and deducing dimensional quantities like covariances or mean cubes by reference to definitive measurements of intensity done once for all.

8.5. Arrangements of Apparatus

In Figs. 47–51 are shown some block diagrams of the arrangement of apparatus needed for some of the more common measurements described in Chapter 2. In the later diagrams, we have omitted the actual anemometer or other transducer and started the block diagrams at the point or points at which electrical signals proportional to the fluctuating quantities are available: the block diagrams are terminated at the point at which a voltage whose d.c. component is proportional to the required mean value is obtained. This voltage is then to be fed into an integrator or an averaging voltmeter. In the case of constant current apparatus, the whole chain can be calibrated in one by inserting a small

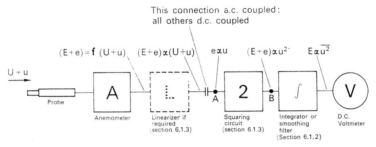

E and e denote mean and fluctuating voltages respectively (different at each point)

A is the point at which the voltage is proportional to the velocity fluctuation

B is the point at which the voltage is proportional to the instantaneous square

FIG. 47. Arrangement of apparatus for measurement of $\overline{u^2}$.

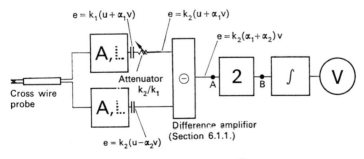

FIG. 48. Arrangement for $\overline{v^2}$.

d.c. signal across the wire terminals (in the absence of the wire). In the case of constant temperature apparatus, it is usually necessary to calibrate the anemometer and the processing equipment separately, the break being made at the output of the anemometer or at the output of the linearizer if one is used. Calibration with an a.c. signal is usually straightforward, but the following points should be remembered.

(1) The gain of the compensating circuit of a hot-wire anemometer must be measured at, or referred to, low frequency since it is the low-frequency gain that is equal to the effective gain of the hot wire and compensator combined.

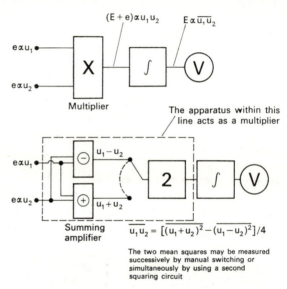

Fig. 49. Alternative arrangements for correlations.

(2) The calibration of squaring circuits or multipliers should be expressed in the form $e_{out} = k e_{in}^2$, where the dimensions of k are volts per volt^2. For instance, if the input is 0·1 V r.m.s. the numerical value of k is 100 times the mean d.c. output voltage.

(3) Apparatus for measuring mean cubes or triple products can still be calibrated by using a sinusoidal signal by rearranging it to form a

fourth power as shown in Fig. 51(b). The required calibration can then be obtained from

$$\frac{\overline{e_1^4}}{(\overline{e_1^2})^2} - 1 = 0\cdot5 = \left(\frac{E}{k_2 k_1^2}\right)\bigg/\left(\frac{E^1}{k_2}\right)^2$$

(E^1 being the mean output of the second squaring circuit when used to square the input signal) giving

$$\frac{\sqrt{k_2}}{k_1} = \frac{E^1}{\sqrt{(2E)}}.$$

In the mean-cube configuration, the skewness coefficient of the turbulence signal, $\overline{u^3}/(\overline{u^2})^{3/2}$, is $(E/(E^1)^{3/2})\sqrt{(k_2)}/k_1$.

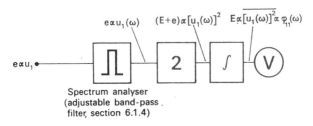

(a) Intensity spectrum

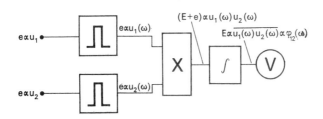

(b) Cross-power spectrum

FIG. 50. Arrangements for spectra.

When dealing with complicated apparatus it is usually a help to draw out a block diagram and to arrange the apparatus in roughly the same physical layout. It will usually be advisable to label at least some of the controls and connections. Connection changes required during the run should be made by switches and not by plugs, although it is not usually worth while to install all the switches required for calibration. Especially with new apparatus, it is sometimes a help to cover with adhesive tape the controls that will not be touched during the course of the experiment (usually at least 80 per cent of the total).

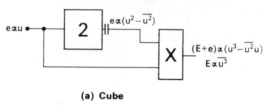

(a) Cube

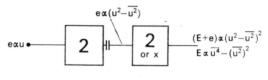

(b) Fourth power

FIG. 51. Arrangement for cubes and fourth powers.

8.6. Distortion of the Flow by the Presence of the Probe

Turbulence-measuring probes need stronger supports than probes for mean flow measurement, because vibration must be avoided. Therefore, they are liable to cause distortion of the mean flow and hence of the turbulence. This large-scale interference with the flow must be distinguished from local interference with the flow round the probe, which depends primarily on the probe and not the flow. Large-scale interference is most troublesome in flows with separation, because a change in separation position, caused by the wake of the probe or its support, may alter the velocity distribution upstream of separation: worst of all is a recirculating flow in which the probe wake may form a closed loop. A general test for interference with the flow is to measure some sensitive quantity (such as the position of a separation line) with and without the probe: flow visualization may be helpful if a sufficiently sensitive technique can be used. Alternatively the probe can be replaced by another with a *larger* body and support: if both probes give the same answer all is probably well. In extreme cases, such as transitional flows, any object in the flow will change it appreciably: one should then ask oneself if experiments on a flow that is so sensitive to its boundary conditions are likely to be of any use outside that particular experimental situation. If the answer is yes, one must use a laser anemometer or some other optical method.

Tritton[70] has shown that severe interference may occur between two probes used for correlation measurements, even if one probe does not lie directly in the wake of the other.

APPENDIX 1

The Equations of Motion[5]

ALL these are "transport" equations (see Sections 1.1 and 2.2, and
Fig. 10) equating the net rate of outward transfer of some quantity Q
through the faces of an infinitesimal control volume to the net sources
of Q within the control volume. Q may be transported in or out of the
volume by the mean flow or by turbulent diffusion (including pressure
fluctuations) or molecular diffusion. It may be produced within the
volume by "external" agencies like body forces or pressure gradients,
or by exchange with another quantity (e.g. production of turbulent
energy at the expense of mean flow kinetic energy). It may be destroyed
either by further exchange (e.g. dissipation of turbulent kinetic energy
into thermal internal energy) or absolutely (e.g. destruction of temperature fluctuations by conductivity). Note that molecular processes
can *either* transport a quantity or dissipate it. The equations for Q are
all based on the Navier–Stokes equations (where $Q \equiv$ momentum and
the conservation principle is Newton's second law of motion) or on
the heat transfer equation (where $Q \equiv$ internal energy or enthalpy and
the conservation principle is the first law of thermodynamics). Equations
for other physical quantities such as kinetic energy or entropy are
obtained by multiplying these equations by powers of velocity or temperature. The equation of conservation of mass stands by itself but is
frequently used to simplify the other conservation equations. Compressibility complicates the equations without introducing new principles
(except for the appearance of "compression work" terms in the energy
equations): here we give only the *incompressible* equations; an exhaustive
treatment of the compressible turbulence equations is given by Favre[71]
and the compressible Navier–Stokes equations are derived by Schlichting[11].

186

Writing U, V and W for the instantaneous velocity components in the x, y and z directions, the **Navier–Stokes equation** for the acceleration of a fluid element in the x direction is

$$\frac{\partial U}{\partial t} + U\frac{\partial U}{\partial x} + V\frac{\partial U}{\partial y} + W\frac{\partial U}{\partial z}$$

$$= -\frac{1}{\varrho}\frac{\partial p}{\partial x} + \nu\left(\frac{\partial^2 U}{\partial x^2} + \frac{\partial^2 U}{\partial y^2} + \frac{\partial^2 U}{\partial z^2}\right), \qquad (1a)$$

from which equations (1b) and (1c), for V and W, can be obtained by changing the symbols in cyclic order. It is a useful shorthand to write the last group of terms as $\nu\,\nabla^2 U$.

The mass conservation or **continuity equation** is

$$\frac{\partial U}{\partial x} + \frac{\partial V}{\partial y} + \frac{\partial W}{\partial z} = 0. \qquad (2)$$

The **enthalpy conservation equation** for a constant-property fluid, writing $c_p\Theta$ for the instantaneous enthalpy, is

$$\frac{\partial \Theta}{\partial t} + U\frac{\partial \Theta}{\partial x} + V\frac{\partial \Theta}{\partial y} + W\frac{\partial \Theta}{\partial z} = \frac{k}{\varrho c_p}\nabla^2\Theta. \qquad (3)$$

In turbulent flow, the velocity components, the pressure and the temperature have mean and fluctuating parts. We can distinguish the two by averaging either over time or over successive repetitions of the experiment ("ensemble average"), the latter being necessary if the "mean" flow is changing with time as in the case of an oscillating aerofoil. In either case we write the mean velocity components as U, V, W and the fluctuations as u, v, w; for pressure we use p, p' and for temperature T, θ to avoid the symbols P, t which have other meanings. We use q^2 as an abbreviation for $u^2 + v^2 + w^2$. Inserting $U + u$ instead of U, and so on, the continuity equation (2) becomes

$$\frac{\partial U}{\partial x} + \frac{\partial V}{\partial y} + \frac{\partial W}{\partial z} + \frac{\partial u}{\partial x} + \frac{\partial v}{\partial y} + \frac{\partial w}{\partial z} = 0. \qquad (4)$$

Taking the mean (time average or ensemble average) we have

$$\frac{\partial U}{\partial x} + \frac{\partial V}{\partial y} + \frac{\partial W}{\partial z} = 0 \qquad (5)$$

and, subtracting this from the instantaneous equation (4), we have

$$\frac{\partial u}{\partial x} + \frac{\partial v}{\partial y} + \frac{\partial w}{\partial z} = 0 \tag{6}$$

as well. Thus the same equation applies to the mean and the fluctuating parts: this is because no products of velocities appear. A product such as $(U + u)(V + v)$ has a mean part $UV + \overline{uv}$, using the overbar to denote the mean part of a fluctuating quantity, and a fluctuating part (i.e. the part with zero mean) $Uv + uV + uv - \overline{uv}$: clearly, products have rather complicated mean and fluctuating parts. If we insert $U + u$ instead of U, and so on, in eqn. (1) and take the mean we have

$$\frac{\partial U}{\partial t} + U\frac{\partial U}{\partial x} + V\frac{\partial U}{\partial y} + W\frac{\partial U}{\partial z} + \overline{u\frac{\partial u}{\partial x}} + \overline{v\frac{\partial u}{\partial y}} + \overline{w\frac{\partial u}{\partial z}}$$

$$= -\frac{1}{\varrho}\frac{\partial \bar{p}}{\partial x} + \nu \nabla^2 U. \tag{7}$$

Again, equations for V and W can be obtained by cyclic interchange. Multiplying (6) by u, adding the mean to (7) and using the rule for differentiation of a product, e.g. $\partial \overline{uv}/\partial y = \overline{u\,\partial v/\partial y} + \overline{v\,\partial u/\partial y}$, we get the **mean momentum equation**

$$\frac{\partial U}{\partial t} + U\frac{\partial U}{\partial x} + V\frac{\partial U}{\partial y} + W\frac{\partial U}{\partial z}$$

$$= -\frac{1}{\varrho}\frac{\partial \bar{p}}{\partial x} - \left(\frac{\overline{\partial u^2}}{\partial x} + \frac{\overline{\partial uv}}{\partial y} + \frac{\overline{\partial uw}}{\partial z}\right) + \nu \nabla^2 U \tag{8}$$

which is more convenient than (7) and enables us to recognize the mean products of fluctuating quantities as extra apparent stresses per unit mass whose gradients apply forces to an element of fluid. This is why we move these mean products to the right-hand side of the equation to accompany the pressure and viscous force terms, leaving the acceleration on the left: it is another useful shorthand to write the acceleration in the x direction as DU/Dt.

Equation (8) enshrines the difficulty of turbulent flow: second-order mean products of fluctuating quantities appear in the equation for the mean velocity. We can obtain equations for $D\overline{u^2}/Dt$, $D\overline{uv}/Dt$ and so on by going back to (1) and its two companion equations but, as we shall see, these equations contain further unknowns. Inserting $U + u$ for U and so on as before, multiplying (1) by the fluctuating component u and taking the mean we obtain

$$\frac{\partial \tfrac{1}{2}\overline{u^2}}{\partial t} + U\frac{\partial \tfrac{1}{2}\overline{u^2}}{\partial x} + V\frac{\partial \tfrac{1}{2}\overline{u^2}}{\partial y} + W\frac{\partial \tfrac{1}{2}\overline{u^2}}{\partial z} \equiv \frac{D \tfrac{1}{2}\overline{u^2}}{Dt}$$

$$= -\frac{1}{\varrho}\frac{\overline{u\,\partial p'}}{\partial x} + \overline{vu\,\nabla^2 u} - \left(\overline{u^2}\frac{\partial U}{\partial x} + \overline{uv}\frac{\partial U}{\partial y} + \overline{uw}\frac{\partial U}{\partial z}\right)$$

$$- \left(\overline{u^2\frac{\partial u}{\partial x}} + \overline{uv\frac{\partial u}{\partial y}} + \overline{uw\frac{\partial u}{\partial z}}\right). \tag{9}$$

By adding the mean of u^2 times eqn. (6), we can rewrite the last three terms as

$$-\frac{1}{2}\left(\frac{\overline{\partial u^3}}{\partial x} + \frac{\overline{\partial u^2 v}}{\partial y} + \frac{\overline{\partial u^2 w}}{\partial z}\right).$$

The sum of (9) and its two companion equations for $D\tfrac{1}{2}\overline{v^2}/Dt$ and $D\tfrac{1}{2}\overline{w^2}/Dt$ is the **turbulent energy equation** for $D\tfrac{1}{2}\overline{q^2}/Dt$, whose physical significance is discussed in Chapter 2. The sum of the first term on the right of (9) and its companions is

$$-\frac{1}{\varrho}\left(\overline{u\frac{\partial p'}{\partial x}} + \overline{v\frac{\partial p'}{\partial y}} + \overline{w\frac{\partial p'}{\partial z}}\right)$$

or

$$-\frac{1}{\varrho}\left(\frac{\overline{\partial p'u}}{\partial x} + \frac{\overline{\partial p'v}}{\partial y} + \frac{\overline{\partial p'w}}{\partial z}\right) + \frac{1}{\varrho}\overline{p'\left(\frac{\partial u}{\partial x} + \frac{\partial v}{\partial y} + \frac{\partial w}{\partial z}\right)}$$

in which the second group of terms is zero by continuity. The first group represents spatial transport of turbulent energy by the action of pressure gradients. Rewriting the first term on the right of (9) as

$$-\frac{1}{\varrho}\frac{\overline{\partial p'u}}{\partial x} + \frac{1}{\varrho}\overline{p'\frac{\partial u}{\partial x}}$$

we can identify the first part as one of the spatial transport terms: the second part represents a transfer of energy from the u-component fluctuations to—necessarily—the other components, and the sum of this part and its companions in the equations for the other components is necessarily zero. The physical significance of terms in the equations is always made clearer by extracting the spatial transport part, which, as is clear from the control-volume concept, is always the "divergence" of a mean product. As another example of this, the sum of the second term on the right of (9) and its companion equations is

$$\nu\overline{(u\,\nabla^2 u + v\,\nabla^2 v + w\,\nabla^2 w)}$$

which is easily seen, by working backwards from the answer, to be

$$\nu\left(\tfrac{1}{2}\nabla^2\overline{(u^2 + v^2 + w^2)} - \overline{\left(\frac{\partial u}{\partial x}\right)^2} - \overline{\left(\frac{\partial u}{\partial y}\right)^2} - \overline{\left(\frac{\partial u}{\partial z}\right)^2} - \text{similar terms}\right).$$

Clearly the first group, written as $\nu\,\nabla^2\tfrac{1}{2}\overline{q^2}$ for short, is the spatial gradient of a mean product, and is understandable as the viscous diffusion of $\tfrac{1}{2}\overline{q^2}$ just as $\nu\,\nabla^2 U$ is the viscous diffusion of U. However, the remaining terms are not equal to the viscous dissipation of turbulent kinetic energy but differ from it by yet another set of spatial gradient terms like $\partial^2\overline{uv}/\partial x\,\partial y$. The viscous dissipation ε is the mean product of the instantaneous viscous stress and the instantaneous rate of strain, which, unlike the terms in the expression above, involves the products of different velocity components. The full analysis is too cumbersome to reproduce without using tensor notation and interested readers are referred to section 2.7 of ref. 5. The point is not of great importance because the viscous diffusion terms are all negligible except when the Reynolds number of the turbulence is very low (for instance, in the viscous sub-layer): in all other cases

$$-\nu\overline{(u\,\nabla^2 u + v\,\nabla^2 v + w\,\nabla^2 w)} \simeq \overline{\left(\frac{\partial u}{\partial x}\right)^2} + \overline{\left(\frac{\partial u}{\partial y}\right)^2} + \overline{\left(\frac{\partial u}{\partial z}\right)^2}$$

$$+ \overline{\left(\frac{\partial v}{\partial x}\right)^2} + \overline{\left(\frac{\partial v}{\partial y}\right)^2} + \overline{\left(\frac{\partial v}{\partial z}\right)^2}$$

$$+ \overline{\left(\frac{\partial w}{\partial x}\right)^2} + \overline{\left(\frac{\partial w}{\partial y}\right)^2} + \overline{\left(\frac{\partial w}{\partial z}\right)^2} \simeq \varepsilon.$$

The last three terms in (9) represent spatial transport of $\tfrac{1}{2}\overline{u^2}$ by the turbulence, and the three terms involving mean velocity gradients are energy production terms. Fortunately we do not often have to deal with either the $\overline{u^2}$ equation or the turbulent energy equation in full: the version of the latter that is most frequently needed is the much-simplified one that applies to a two-dimensional shear layer obeying the "boundary layer approximation", so that x derivatives are small compared with y derivatives and z derivatives are zero. Further assuming that the Reynolds number is high enough for viscous diffusion of turbulent energy to be negligible, we have

$$\frac{D}{Dt}\tfrac{1}{2}\overline{q^2} = -\overline{uv}\frac{\partial U}{\partial y} - \frac{\partial}{\partial y}\left(\frac{\overline{p'v}}{\varrho} + \tfrac{1}{2}\overline{q^2 v}\right) - \varepsilon.$$

By repeating the derivation of the $D\tfrac{1}{2}\overline{u^2}/Dt$ equation but multiplying (1a) by v instead of u and adding u times (1b), we get an equation for $D\overline{uv}/Dt$, of which the first few terms are

$$\frac{D\overline{uv}}{Dt} = -\frac{1}{\varrho}\left(\frac{\overline{\partial p'\,u}}{\partial y} + \frac{\overline{\partial p'\,v}}{\partial x}\right) + \frac{1}{\varrho}\overline{p'\left(\frac{\partial u}{\partial y} + \frac{\partial v}{\partial x}\right)}$$
$$+ \nu(\overline{u\,\nabla^2 v} + \overline{v\,\nabla^2 u}) + \dots \tag{10}$$

and similarly for the other shear stress components. The second set of terms on the right represents a "scrambling" of the turbulence by pressure fluctuations: usually it tends to make the turbulence more nearly isotropic ($\overline{u^2} = \overline{v^2} = \overline{w^2}$, $\overline{uv} = 0, \dots$).

Equations such as (9) and (10) contain third-order products on the right: D/Dt equations for the latter can be obtained by multiplying (1) by the product of *two* velocities, but the resulting equations contain fourth-order products, and so on.

The pressure fluctuation is a hidden second-order product or, rather, the integral of second-order products over all space. Differentiating (1) with respect to x, and the companion V and W equations with respect to y or z (that is, taking the divergence of the Navier–Stokes equations) we get the **pressure equation**

$$-\frac{1}{\varrho}\nabla^2(\bar{p} + p') = \frac{\partial^2 UV}{\partial x\,\partial y} + \frac{\partial^2 VW}{\partial y\,\partial z} + \frac{\partial^2 WU}{\partial z\,\partial x}$$
$$+ \frac{\partial^2 U^2}{\partial x^2} + \frac{\partial^2 V^2}{\partial y^2} + \frac{\partial^2 W^2}{\partial z^2}$$

where U, V and W are still the *instantaneous* velocities. Note that the viscous stresses do not contribute to the pressure (p. 7). The solution of Poisson's equation, $\nabla^2 p = f$, say, is

$$4\pi p(\mathbf{x}) = \iiint \frac{f(\mathbf{x}')}{|\mathbf{x} - \mathbf{x}'|} \, dx' \, dy' \, dz'$$

where the integral is over all the values of the position vector $\mathbf{x}'$ that contribute to f, and $|\mathbf{x} - \mathbf{x}'|^2 = (x - x')^2 + (y - y')^2 + (z - z')^2$. It hardly needs saying that this "solution" is not much help when f comes from a random velocity field, but $\overline{p'^2}$ can be expressed as integrals of correlations.

The **mean enthalpy equation**, analogous to eqn. (8), is

$$\frac{\partial T}{\partial t} + U\frac{\partial T}{\partial x} + V\frac{\partial T}{\partial y} + W\frac{\partial T}{\partial z} = \frac{k}{\varrho c_p}\nabla^2 T - \left(\frac{\partial \overline{u\,\theta}}{\partial x} + \frac{\partial \overline{v\,\theta}}{\partial y} + \frac{\partial \overline{w\,\theta}}{\partial z}\right) \tag{11}$$

where T and θ are the mean and fluctuating parts of the temperature, so that $\Theta = T + \theta$. Multiplying (3) by θ and taking the mean we obtain an equation for $\overline{\theta^2}$ analogous to the turbulent energy equation for $\overline{q^2}$:

$$\frac{D\,\tfrac{1}{2}\overline{\theta^2}}{Dt} = \frac{k}{\varrho c_p}\overline{\theta\,\nabla^2\theta} - \left(\overline{u\theta}\frac{\partial T}{\partial x} + \overline{v\theta}\frac{\partial T}{\partial y} + \overline{w\theta}\frac{\partial T}{\partial z}\right)$$

$$-\frac{1}{2}\left(\frac{\partial \overline{u\,\theta^2}}{\partial x} + \frac{\partial \overline{v\,\theta^2}}{\partial y} + \frac{\partial \overline{w\,\theta^2}}{\partial z}\right). \tag{12}$$

$\overline{\theta^2}$ is *not* an energy quantity: kinetic energy per unit mass is $\tfrac{1}{2}(\text{velocity})^2$, so that total kinetic energy is $\tfrac{1}{2}(U^2 + V^2 + W^2 + \overline{q^2})$, but enthalpy per unit mass is $c_p \times$ temperature $= c_p(T + \theta)$ so that total mean enthalpy is $c_p T$. The first term on the right of (12) represents conductive diffusion and "dissipation" of $\overline{\theta^2}$. Note that unlike dissipation of turbulent energy into heat, dissipation of $\overline{\theta^2}$ does not produce another form of energy: it just represents annihilation of temperature fluctuations by conductivity, as in a solid.

The second group of terms represent "production" of $\overline{\theta^2}$ by the effect of turbulent fluctuations on the mean temperature gradient, and the third group represents spatial transport of $\overline{\theta^2}$ by the turbulent motion.

The pressure does not appear in the temperature equations in an incompressible fluid. However, we do find pressure terms in the equation for the temperature–velocity correlations which produce turbulent heat transfer. Multiplying (1b) by θ and (3) by v, adding and taking the mean we get, using the continuity equation,

$$\frac{D\overline{v\theta}}{Dt} = -\frac{1}{\varrho}\,\overline{\theta\frac{\partial p'}{\partial y}} + \frac{k}{\varrho c_p}\,\overline{v\,\nabla^2\theta} + \overline{v\theta\,\nabla^2 v}$$

$$-\left(\overline{u\theta}\frac{\partial V}{\partial x} + \overline{v\theta}\frac{\partial V}{\partial y} + \overline{w\theta}\frac{\partial V}{\partial z}\right)$$

$$-\left(\overline{uv}\frac{\partial T}{\partial x} + \overline{v^2}\frac{\partial T}{\partial y} + \overline{wv}\frac{\partial T}{\partial z}\right) \tag{13}$$

$$-\left(\frac{\partial \overline{u\,v\theta}}{\partial x} + \frac{\partial \overline{v^2\,\theta}}{\partial y} + \frac{\partial \overline{w\,v\theta}}{\partial z}\right).$$

The main qualitative difference between this and the equation for $D\overline{uv}/Dt$ is that there are two sets of molecular (conductive and viscous terms) and two sets of "production" terms.

At present, these equations are little better than a support for our physical ideas about turbulence, but one must remember that (10) and (13), and the equations for the other Reynolds stresses and turbulent heat-transfer rates, are the equations we are *trying* to solve whenever we make hypotheses about turbulent transfer processes. Therefore when we measure turbulence we are—explicitly or implicitly—investigating the terms on the right-hand sides of these equations. Of course, it is only in fundamental research work that one would think of measuring these terms directly, but even in direct work on industrial problems it is worth bearing in mind the physical significance of these terms, which represent the birth, migration and decay of turbulent eddies.

APPENDIX 2

Turbulence Research

THE main application of turbulence research is to the prediction of mass, heat and momentum transfer in shear flows: but research projects range from direct attacks on this problem, through measurements of mean or fluctuating quantities in laboratory test rigs, to analytical or numerical studies of the Navier–Stokes equations or approximations thereto (see the "Further Reading" list, p. 208).

A2.1. Numerical Solutions for Instantaneous Quantities

At present, solutions of the complete time-dependent Navier–Stokes equations are limited to very low Reynolds numbers because only in this case is the range of length scales small enough for present-day computer storage: if the largest scale is l and the smallest η, then, to give an adequate sample of the velocity field, points must be spaced at a distance ("grid length") rather less than η over a volume much greater than l^3 so that the number of points (and equations!) required is much greater than l^3/η^3 which is of the order of $(\sqrt{\overline{u^2}}l/\nu)^{9/4}$. However, if the universality of the small scale motion (2.6.2) is accepted, the spacing of points need be no smaller than, say, $0{\cdot}01l$, providing that the transfer of energy to eddies smaller than those explicitly represented is simulated by an eddy viscosity or similar means. This approach offers the only hope of direct calculations at Reynolds numbers of practical interest, but even such a restricted calculation is barely within the limits of foreseeable computers. Deardorff[72] has presented calculations for a turbulent channel flow which, although obviously suffering from numerical inaccuracy due to too sparse a distribution of calculation points, indicate that the technique is feasible. Schumann[73] has used similar numerical "measurements" to calibrate an empirical calculation method.

194

Meteorologists studying the general circulation of the atmosphere have done extensive numerical calculations using horizontal grid lengths of the order of 100 km at up to ten different heights. On the global scale, the motion is controlled mainly by Coriolis forces, the energy coming from the sun, either directly or via the evaporation or condensation of water vapour: turbulent transfer can be adequately represented by quite crude eddy diffusivities. Therefore the numerical problems are closely the same as in calculations like those of Deardorff or Schumann.

Although there is no positive reason to suppose that any simplified version of the Navier–Stokes equations can adequately represent turbulence many workers have sought such a simplification. The incompressible equations themselves could scarcely be simpler (see Chapter 1) and the usual approach is to simplify the equations for the development in time of the multi-point mean products of various orders, i.e. the equations with left-hand sides of the form,

$$\frac{D}{Dt}\left(\overline{u_i(\mathbf{x})\, u_j(\mathbf{x} + \mathbf{r})\, u_k(\mathbf{x} + \mathbf{r'})\ldots}\right)$$

where u_i, u_j, u_k are any components of the velocity fluctuations. These equations are obtained by taking the Navier–Stokes equation for each velocity in turn, multiplying by the product of all the other velocities, and adding up all such equations (see Appendix 1 for simple examples). Since the right-hand side of an equation for the mean product of n velocities contains the product of $n + 1$ velocities, an infinite number of such equations is needed to represent the mean products completely. Truncation of this infinite set by, say, setting $(n + 1)$-order products equal to zero, amounts to a simplification of the Navier–Stokes equations: we have to solve only, say, half a dozen mean-product equations instead of satisfying the Navier–Stokes equations at more than l^3/η^3 points. However, neglect of the third-order products amounts to a restriction to Reynolds numbers much less than unity, neglect of the fourth-order "cumulants" (the fourth-order mean products less their values for widely separated points) leads to negative spectral density in numerical calculations, and there is no guarantee that higher truncations will be an improvement. There are several approximations of these equations (or the corresponding spectrum equations) that do not amount to simple truncation. One of the most appealing is the "Lagrangian History Direct Interaction" (LHDI) approximation, developed by

Kraichnan[74] after ten years of work on the subject. Direct Interaction is a means of approximating the exchange of energy between groups of three different wave numbers which appears in the energy spectrum equation, ignoring the effect of wave numbers outside a given group. A defect of this physically plausible approximation is that in reality a given eddy is convected by the combined motion of all other eddies, so that (roughly speaking) the eddy must be studied in a coordinate system moving with it. The Lagrangian modification of Kraichnan's theory approximates this: unfortunately, although Direct Interaction is a rigorous first approximation to the energy transfer, the Lagrangian modification is arbitrary and there is no obvious way of improving it. Another arbitrary but successful modification is Kraichnan's Test Field Model[74].

A2.2. Prediction Methods for Shear Flows[75]

There is as yet little contact between the work described in the last section, whose object is to predict the complete energy spectrum or the multipoint covariances, and the engineering problem of predicting Reynolds stresses (i.e. one-point covariances) which is normally attacked at a much lower level. For slowly changing flows, in which the transport terms in the Reynolds stress equations are negligible compared with the source terms (generation of Reynolds stress by interaction with the mean velocity gradients, and destruction by turbulent "scrambling" or viscous dissipation) so that the latter are equal and opposite, a close relation between the Reynolds stresses and the mean velocity gradients may be expected. By analogy with laminar flow, in which the stresses are directly proportional to the velocity gradients, we define an "eddy viscosity" μ_e, equal to (shear stress) $\div$ (velocity gradient) in a simple shear flow: if the above-mentioned close relation really exists, μ_e will be related to suitable length and velocity scales of the flow, say

$$\mu_e = \varrho u_\tau \, \delta f \left(\frac{y}{\delta} \right)$$

in a boundary layer, where f is an empirical function. This is indeed a fair approximation for slowly changing shear layers (Section 3.4). Also, for flows with similar velocity-defect profiles[5], whose only relevant length scales are u_τ and δ or their equivalents, the above eddy viscosity formula follows from dimensional analysis (in this case the generation and destruction terms are not necessarily equal but their ratio is a function of y/δ only). Finally, the formula $\mu_e = \varrho(\tau/\varrho)^{\frac{1}{2}} y$. constant is accurate in the inner layer of any wall layer, again for dimensional reasons (replace u_τ by $(\tau/\varrho)^{\frac{1}{2}}$ and δ by y in the above). Simple eddy viscosity formulae are *not* valid in the outer layer of rapidly changing wall flows or in rapidly changing free flows, and calculation methods based on such formulae are limited in application, although they have sufficed up to now for many engineering purposes, mainly because the inner layer covers a large fraction of the total velocity change between surface and external stream.

The next stage in sophistication is to relate the Reynolds stresses to the mean velocity gradients by approximate forms of one or more of the Reynolds stress "conservation" equations: here, the mean velocity gradients affect only the *rates of change* of the Reynolds stresses [the left-hand sides of eqns. (9) or (10) of Appendix 1] and do not directly determine the absolute values of the Reynolds stresses. Empirical information is required to express the destruction terms and the turbulent transport terms as functions of the Reynolds stresses so that the number of unknowns shall not exceed the number of equations. It is at this level that turbulence theory meets turbulence measurement, because direct measurements of the terms in the Reynolds stress equations are needed if their approximate representations as functions of Reynolds stress are to be based on anything better than trial-and-error comparison of mean-flow predictions with experiment. The limitations of turbulence measurement techniques limit the equations we can handle: in particular, the Reynolds stress equations all have terms containing the pressure fluctuation and our treatment of these is uncertain.

In general the approximations to terms in the Reynolds stress equations will contain length scales as well as the Reynolds stresses themselves. Providing that the shear layer is not changing really rapidly it is reasonable to assume that these length scales have the form $\delta f(y/\delta)$ where δ is the width of the turbulent region (which is of course the same as or at least closely related to, the width of the shear layer as usually defined). However, in rapidly distorted flows we need one or more differential equations for length scales: such equations can be obtained by integrating a two-point mean product equation (Section A2.1) over all values of the distance between the two points, so that the left-hand side is the rate of change of a one-point mean product times a length scale (a general form of the integral scale defined in Section 2.6.2). The terms in these equations are more complicated than those in the Reynolds stress equations, and existing measurements give little help in approximating them by functions of the Reynolds stresses and the length scale in order to close the set of equations[76].

When substituted into the mean momentum equations, these different assumptions about the turbulent stresses all lead to partial differential equations for the mean velocity components. Now partial differential equations of the boundary layer type can be solved, at least in two dimensions, by representing the variables as analytic functions of one coordinate (say y), leading to ordinary differential equations for the x-dependence of the coefficients of the analytic functions or, equivalently, for the x-dependence of θ, $H \equiv \delta^*/\theta$, and possibly other boundary-layer integral parameters. In the past, such ordinary differential equations have sometimes been derived directly from physical arguments or correlations of mean-flow data, without appeal to hypotheses about the Reynolds stresses at a point. Since the behaviour of the Reynolds stresses depends strongly on y, its representation by direct assumptions about integral parameters is hazardous. However, there is no physical objection to the purely mathematical process of passing from a small number of partial differential equations to a large number of ordinary differential equations for integral parameters: the two systems become equivalent as the number of integral parameters tends to infinity. If the integral parameters are ill-chosen the ordinary differential equations may become ill-conditioned or indeterminate at some values of x: this difficulty becomes more acute as the number of parameters increases, so that integral methods are likely to survive only to provide quick, rough answers for simple flows.

A2.3. Prediction of Turbulent Flows other than Thin Shear Layers

Since thin shear layers (boundary layers, pipe flows, wakes and jets) are much simpler than flows about bluff bodies and in ducts of complicated cross-section, it is not surprising that the greatest theoretical progress has been made in thin shear layers. The mathematical complication of three-dimensional and separated flows is considerable, and even for laminar flows a completely satisfactory numerical treatment is not available: however, the main obstacle to treatment of complicated turbulent flows is lack of physical understanding of the turbulence processes. Not only are there only a few isolated measurements of spectra, correlations and the like, but there are scarcely any mean flow measurements of sufficient detail to serve as test cases for prediction methods. Indeed, the description "turbulent flows other than thin shear layers" covers such a vast range of flows that it is difficult even to suggest experiments which would furnish test cases of sufficient generality: however, many flows are composed of *fairly* thin shear layers, recognizable as distortions of the classical thin shear layers.

Of course there are many prediction methods with a limited range of validity, for cases such as base flows (for example, the flow behind a bullet), shock-wave/boundary-layer interaction and other flows in which the turbulent region is initially thin and then thickens during interaction with the external stream. However, these methods are often intended to predict only the gross features of the flow (base pressure, pressure rise to separation) and may be little more than data correlations based on theories for thin shear layers. There is one and only one objection to data correlations—one cannot confidently extrapolate them. Since all science rests on data correlations of one sort or another, they clearly have a part to play in turbulence predictions, but because of the wide variety of turbulent flows it is exceptionally difficult to be sure that one has included all the variables in the data correlation, and therefore difficult to decide when the correlation is being extrapolated.

A2.4. Experimental Work

Only recently, with the advent of prediction methods involving hypotheses about turbulence quantities, has it been possible to use turbulence measurements (other than Reynolds stress measurements) directly in flow prediction. It remains to be seen how this will affect experimental work: in the past the accent has been on correlation and spectrum measurements for investigating eddy structure, together with measurements of the turbulent energy balance. Of course, knowledge about eddy structure is a help in formulating hypotheses about the

Reynolds stresses, but it seems likely that there will be a swing towards measurements of the terms in the Reynolds stress equations themselves. In the absence of any reliable method of measuring terms involving pressure fluctuations, it will be necessary to measure all the other terms accurately enough for the pressure terms to be deduced by difference.

Apart from Reynolds stresses and rates of turbulent heat or mass transfer, the only fluctuating quantities of *direct* interest to a designer are fluctuating loads, either small-scale pressure fluctuations on a surface below a shear layer or large-scale buffeting of bodies which have turbulent wakes or which are immersed in a turbulent stream. The length scale or frequency spectrum, as well as the mean-square fluctuating load itself, is of interest in avoiding fatigue damage or the excitation of large oscillations at the natural frequencies of the structure.

Semi-quantitative measurements of turbulence intensity or scale are frequently made in engineering investigations where a rough idea of turbulence behaviour is useful for design purposes or as an aid to understanding the flow patterns. Unfortunately, few of the turbulence investigations made for engineering purposes are ever published and there is no general fund of knowledge about what measurements to make and how to interpret them. What is certain is that a single figure for mean-square u-component intensity tells one little about the flow: much of the contribution may come from frequencies so low as to amount to unsteadiness rather than turbulence (this is especially true of wind tunnel "turbulence" which is mostly at wavelengths longer than the working section). Therefore spectrum measurements, at least, are necessary unless the flow is fairly well understood from previous measurements. Usually the Reynolds stresses (especially the shear stress) or other turbulent transport rates are the quantities of prime interest, and it is probably worth the extra effort to measure, say, $\overline{uv}$ instead of $\overline{u^2}$: large-scale unsteadiness will not usually contribute to $\overline{uv}$, for reasons given in Section 3.3.

Conditional sampling, using the techniques outlined at the end of Chapter 2, has recently become very popular. The cautionary remarks in the last sentence of Chapter 2 still apply.

Notation

A   "Open-loop" or "intrinsic" gain of amplifier [voltage amplification ratio in absence of feedback (6.1.1)]: additive constant in logarithmic velocity profile (3.3)

C   Capacitance (6.1.2)

c   Specific heat (5.7): concentration (3.9, 7.6): speed of light (4.3)

c_p Specific heat at constant pressure (3.9)

d   Diameter (of pipe or of hot-wire)

E   Mean voltage (especially across hot-wire (5.1.3): wave-number-magnitude spectrum (2.6.1)

f   Frequency, Hertz (cycles per second): a function (not necessarily the same function at each time of use)

g   Gravitational acceleration (1.2, 3.7)

H   Heat transfer per unit time (5.1; compare Q)

i   Current: $\sqrt{-1}$

j $\sqrt{-1}$

K   Constant in logarithmic velocity profile (3.3)

k   Wave number magnitude, $\sqrt{(k_1^2 + k_2^2 + k_3^2)}$ (2.6): roughness height (3.3), thermal conductivity

k   Wave number vector, components k_1, k_2, k_3 in x, y, z directions (2.6)

L   Integral length scale $\int_0^\infty R \, dr$, generally $\int_0^\infty R_{11}(r, 0, 0) \, dr_1$ (2.6)

l   Length scale, length of hot-wire

l_c "Cold length" (5.1)

M   Time constant (of hot-wire) (4.2)

m   Mass

P   Probability (2.10)

p   Mean pressure

p'   Pressure fluctuation

Q   Heat transfer per unit area per unit time (3.9; compare H)

q   Resultant velocity fluctuation, $\sqrt{(u^2 + v^2 + w^2)}$ (2.2)

R Electrical resistance (5.1, 6.1): correlation (2.3)

r Correlation separation, components r_1, r_2, r_3 in the x, y, z directions

T Mean temperature (3.9)

T_0 Stagnation temperature (7.2)

t Time

U, V, W Mean velocity components in x, y, z directions

U_c Convection velocity (4.8)

U_∞ Velocity at edge of shear layer (3.4)

u, v, w Velocity fluctuations in x, y, z directions

u_τ Friction velocity, $\sqrt{(\tau_w/\varrho)}$ (3.3)

v Kolmogorov velocity scale (2.6.2)

x, y, z Cartesian coordinate axes, x in general direction of flow, y normal to surface or plane of shear layer. Note that in meteorology z is vertical (3.7)

Z Impedance (6.2)

z Distance along hot-wire from mid-point (5.1)

α Temperature coefficient of resistance (5.1)

γ Intermittency (2.10)

δ Shear layer thickness (3.4)

δ^* Displacement thickness $\int_0^\delta (1 - U/U_\infty)\,dy$ in incompressible flow (3.4)

ε $\varrho\varepsilon$ is rate of viscous dissipation of turbulent kinetic energy per unit volume (2.2). Note that in some books ε is eddy viscosity

η Kolmogorov length scale (2.6.2)

θ Temperature fluctuation (3.9): momentum-deficit thickness, $\int_0^\delta (1 - U/U_\infty)\,U/U_\infty\,dy$ in incompressible flow (3.4)

λ Wavelength, microscale (2.6.2)

μ Molecular viscosity

ν Kinematic viscosity μ/ϱ

ϱ Density

σ_{xx} Tensile stress in x direction $\big\}$ similarly for other suffixes
σ_{xy} Shear stress in x, y-plane

τ Shear stress in x, y-plane (3.3): time delay (2.4)

Φ Three-dimensional wave-number spectral density (2.6)

ϕ	Frequency spectral density (2.5): one-dimensional wave-number spectral density (2.6)
ψ	Yaw angle (5.3)
ω	Radian frequency, angular velocity

Suffixes

a	(Ambient) at fluid temperature
f	Fluid
w	(Wall) surface, or wire
0	Initial or reference conditions (T_0 is stagnation temperature)
τ	Only in $u_\tau \equiv \sqrt{(\tau_w/\varrho)}$

Dimensionless numbers

Gr	Grashof number (5.1)
Kn	Knudsen number (5.1)
M	Mach number (1.7, 7.2)
Nu	Nusselt number (5.1)
Pr	Prandtl number (3.9, 5.1)
Re	Reynolds number (1.6, 3.4, 5.1)
Ri	Richardson number (3.7)
St	Stanton number (3.9, 5.1)

Overbar (e.g. $\bar{f}$); mean value [(1.4), and see Glossary under "Average"]

References

1. ROSENHEAD, L. (Ed.). *Laminar Boundary Layers*, University Press, Oxford (1963), Ch. 1 and Ch. 3, pt. 1; for a description using tensor notation, see BATCHELOR, G. K. *An Introduction to Fluid Dynamics*, University Press, Cambridge (1967).
2. BATCHELOR, G. K. *The Theory of Homogeneous Turbulence*, University Press, Cambridge (1956).
3. CORRSIN, S. and KARWEIT, M. Fluid line growth in grid-generated isotropic turbulence, *J. Fluid Mech.* **39**, 87 (1969).
4. LUMLEY, J. L. and PANOFSKY, H. A. *The Structure of Atmospheric Turbulence*, Wiley, New York (1964).
5. TOWNSEND, A. A. *The Structure of Turbulent Shear Flow*, University Press, Cambridge (1956).
6. ROSE, W. G. Results of an attempt to generate a homogeneous turbulent shear flow, *J. Fluid Mech.* **25**, 97 (1966); see also *J. fluid Mech.* **44**, 767 (1970).
7. KORKEGI, R. H. and BRIGGS, R. A. On compressible turbulent plane Couette flow, *A.I.A.A. Journal* **6**, 742 (1968).
8. KLINE, S. J., REYNOLDS, W. C., SCHRAUB, F. A. and RUNSTADLER, P. W. *J. Fluid Mech.* **30**, 741 (1967).
9. CORINO, E. R. and BRODKEY, R. S. A visual investigation of the wall region in turbulent flow, *J. Fluid Mech.* **37**, 1 (1969).
10. COLES, D. E. and HIRST, E. A. (Eds.). *Proceedings: Computation of Turbulent Boundary Layers—1968 AFOSR-IFP-Stanford Conference*, Vol. 2, Thermosciences Division, Stanford Univ. (1969).
11. SCHLICHTING, H. *Boundary Layer Theory*, Pergamon, Oxford (1968).
12. ROTTA, J. C. Turbulent boundary layers in incompressible flow, *Progress in Aero. Sci.* **2**, 1 (1962).
13. BRAGG, G. M. The turbulent boundary layer in a corner, *J. Fluid Mech.* **36**, 485 (1969).
14. CALVERT, J. R. Experiments on the low-speed flow past cones, *J. Fluid Mech.* **27**, 273 (1967).
15. RUDD, M. J. Measurements made on a drag reducing solution with a laser velocimeter, *Nature* **224**, 587 (1969).
16. SANDBORN, V. A. *Resistance-temperature Transducers*, to be published.
17. FERRISS, D. H. Measurements of the free stream turbulence in the R.A.E. Bedford 13 ft × 9 ft wind tunnel. *Aero. Research Council C.P.* 719 (1963).
18. TROLINGER, J. D. Laser applications in flow diagnostics, *AGARDOgraph* No. 186 (1974).
19. FISHER, M. J. and KRAUSE, F. R. The crossed beam correlation technique, *J. Fluid Mech.* **28**, 705 (1967).

20. VASUDEVAN, M. S. and JANSSON, R. E. W. Optical methods of measuring turbulence, *Southampton Univ. A.A.S.U. Rep.* 288 (1969).
21. REID, A. M. Turbulence measurement using a glow discharge as an anemometer. *R. and D.*, No. 6, p. 69 (Feb. 1962).
22. FRANZEN, B., FUCKS, W. and SCHMITZ, G. Koronaanemometer zur Messung von Turbulenzkomponenten, *Zeit. für Flugwiss.* **9**, 347 (1961).
23. GASTER, M. A new technique for the measurement of low fluid velocities, *J. Fluid Mech.* **20**, 183 (1964).
24. BAUER, A. B. Direct measurement of velocity by hot-wire anemometry, *A.I.A.A. Journal* **3**, 1189 (1965).
25. TOMBACH, I. H. Velocity measurements with a new probe in inhomogeneous turbulent jets. Ph. D. Thesis, California Institute of Technology (1969).
26. BRADBURY, L. J. S. and CASTRO, I. P. A pulsed wire technique for velocity measurements in highly-turbulent flows. *J. Fluid Mech.* **49**, 657 (1971).
27. BRADSHAW, P. *Experimental Fluid Mechanics*, Pergamon, Oxford (1969).
28. BRADSHAW, P. and GOODMAN, D. G. The effect of turbulence on static-pressure tubes, *Aero. Research Council R. & M.* 3527 (1968).
29. GEIB, F. E. Measurements of the effect of transducer size on the resolution of boundary-layer pressure fluctuations, *J. Acoust. Soc. America,* **46**, 253 (1969).
30. COLLIS, D. C. and WILLIAMS, M. J. Two-dimensional convection from heated wires at low Reynolds numbers, *J. Fluid Mech.* **6**, 357 (1959).
31. CORRSIN, S. Turbulence: experimental methods. *Handbuch der Physik*, Vol. 8, Part 2, Springer, Berlin (1963).
32. BELLHOUSE, B. J. and SCHULTZ, D. L. The measurement of fluctuating skin friction in air with heated thin-film gauges, *J. Fluid Mech.* **32**, 675 (1968).
33. BELLHOUSE, B. J. and RASMUSSEN, C. G. Low-frequency characteristics of hot-film anemometers, *DISA Information* No. 6 (1968).
34. BOURKE, P. J., PULLING, D. J., GILL, L. E. and DENTON, W. H. The measurement of turbulent velocity fluctuations and turbulent temperature fluctuations in the supercritical region by a hot-wire anemometer and a "cold" wire resistance thermometer. *Proc. Instn. Mech. Engrs.* **182**, Part 31 (1967–8).
35. DAVIES, P. O. A. L. and FISHER, M. J. Heat transfer from electrically heated cylinders, *Proc. Roy. Soc.* A **280**, 486 (1964).
36. BEARMAN, P. W. Corrections for the affect of ambient temperature drift on hot wire measurements in incompressible flow, *DISA Information* No. 11 (1971).
37. DAHM, M. and RASMUSSEN, C. G. Effect of wire mounting systems on hot-wire probe characteristics, *DISA Information* No. 7 (1969).
38. COMTE-BELLOT, G. Contribution à l'étude de la turbulence de conduite, Ph. D. Thesis, University of Grenoble (1963): translated as Aero. Research Council Paper No. 31609 (1969).
39. BRADSHAW, P. "Inactive" motion and pressure fluctuations in turbulent boundary layers, *J. Fluid Mech.* **30**, 241 (1967).
40. CHAMPAGNE, F. H., SCHLEICHER, C. A. and WEHRMANN, O. H. Turbulence measurements with inclined hot-wires, Parts 1 and 2, *J. Fluid Mech.* **28**, 153 (1967).
41. FRIEHE, C. A. and SCHWARTZ, W. H. Deviation from the cosine law for yawed cylindrical anemometer sensors. *Trans. A.S.M.E.* **35E**, 655 (1968).

42. RASMUSSEN, C. G. The air bubble problem in hot-film anemometry, *DISA Information* No. 5 (1967).

43. WYNGAARD, J. C. Measurements of small-scale turbulence structure with hot wires, *J. Sci. Instrum.*, Ser. 2, **1**, 1105 (1968).

44. CLAYTON, G. B. Operational amplifiers, *Wireless World*, Feb.–Sept. 1969 (8 parts).

45. GIBSON, C. H. Digital techniques in turbulence research, *AGARDograph* No. 174 (1973).

46. BOURKE, P. J. and PULLING, D. J. A turbulent heat flux meter and some measurements of turbulence in air flow through a heated pipe, *International J. Heat & Mass Transf.* **13**, 1331 (1970).

47. MORKOVIN, M. V. Fluctuations and hot-wire anemometry in compressible flows, *AGARDograph* 24 (1956).

48. LAUFER, J. and McLELLAN, R. Measurement of heat transfer from fine wires in supersonic flow, *J. Fluid Mech.* **1**, 276 (1956).

49. BOLTZ, F. W. Hot-wire heat loss characteristics and anemometry in subsonic continuum and slip flow. *N.A.S.A. TN* D-773 (1961).

50. VREBALOVICH, T. Applications of hot-wire techniques in unsteady compressible flows, *Symposium on Measurement in Unsteady Flow*, A.S.M.E., New York (1962),

51. ARYA, S. P. S. and PLATE, E. J. Hot-wire measurements in non-isothermal flow, *Instruments and Control Systems* **42**, 87 (1969).

52. CORRSIN, S. Extended applications of the hot-wire anemometer, *N.A.C.A.* TN 1864 (1949).

53. CONGER, W. L. The measurement of concentration fluctuations in the mixing of two gases by hot-wire anemometer techniques, Ph. D. Thesis, Penn. State Univ., Univ. Microfilms Inc. no. 66–255 (1965).

54. WAY, J. and LIBBY, P. A. Hot-wire probes for measuring velocity and concentration in helium–air mixtures, *A.I.A.A. Journal* **8**, 976 (1970).

55. GIBSON, C. H. and SCHWARZ, W. H. The universal equilibrium spectra of turbulent velocity and scalar fields, *J. Fluid Mech.* **16**, 365 (1963).

56. NYE, J. O. and BRODKEY, R. S. Light probe for measurement of turbulent concentration fluctuations, *Rev. Sci. Instrum.* **38**, 26 (1967).

57. BECKER, H. A., HOTTEL, H. C. and WILLIAMS, G. C. On the light-scatter technique for the study of turbulence and mixing, *J. Fluid Mech.* **30**, 259 (1967).

58. FIEDLER, H. and HEAD, M. R. Intermittency measurements in the turbulent boundary layer, *J. Fluid Mech.* **25**, 719 (1966).

59. FUJITA, H. and KOVASZNAY, L. S. G. Measurement of Reynolds stress by a single rotated hot-wire anemometer, *Rev. Sci. Instrum.* **39**, 1351 (1968).

60. BELLHOUSE, B. J. and BELLHOUSE, F. H. Thin-film gauges for the measurement of velocity or skin friction in air, water or blood, *J. Sci. Instrum (J. Phys. E)* **1**, 1211 (1968).

61. DRING, E. P. and GEBHART, B. Hot-wire anemometer calibration for measurement at very low velocity, *Trans. A.S.M.E.* **91C**, 241 (1969).

62. MALCOLM, D. G. Some aspects of turbulence measurement in liquid mercury using cylindrical quartz-insulated hot-film sensors, *J. Fluid Mech.* **37**, 701 (1969).

63. Hoff, M. Hot-film anemometry in liquid mercury, *Instruments and Control Systems* **42**, 832 (1969).

64. Poreh, M., Landa, I. and Kidron, I. On the measurement of low-level turbulence using a pair of sensors, *Israel J. Technol.* **7**, 55 (1969).

65. Guitton, D. E. Correction of hot-wire data for high intensity turbulence, longitudinal cooling and probe interference, *McGill Univ., Mech. Engng. Res. Lab. Rep.* 68–6 (1968).

66. Rose, W. G. Some corrections to the linearized response of a constant-temperature hot-wire anemometer operated in a low speed flow, *Trans. A.S.M.E.* **29E**, 554 (1962).

67. Richardson, E. V., and McQuivey, R. S. Measurement of turbulence in water, *J. Hydr. Div., A.S.C.E.*, **94**, HY2, 411 (1968).

68. Resch, F. and Coantic, M. Étude sur le fil chaud et le film chaud dans l'eau, *Houille Blanche* **24**, 151 (1969).

69. Astarita, G. and Nicodemo, L. Behaviour of velocity probes in viscoelastic dilute polymer solutions, *Indust. and Engng. Chem. Fundamentals* **8**, 582 (1969).

70. Tritton, D. J. Note on the effect of a nearby obstacle on turbulent intensity in a boundary layer, *J. Fluid Mech.* **28**, 433 (1967).

71. Favre, A. J. Équations des gaz turbulents compressibles, *J. de Méc.* **4**, 361 and 391 (1965).

72. Deardorff, J. W. A numerical study of three dimensional turbulent channel term at large Reynolds numbers, *J. Fluid Mech.* **41**, 453 (1970).

73. Schumann, U. Procedure for the direct numerical simulation of turbulent flows. Kernforschungszentrum Karlsruhe, KFK-1854 (1973).

74. Leslie, D. C. Developments in the Theory of Turbulence, Clarendon Press, Oxford (1973).

75. Kline, S. J., Morkovin, M. V., Sovran, G. and Cockrell D. J. (Eds.). Proceedings, *Computation of Turbulent Boundary Layers—1968 AFOSR-IFP-Stanford Conference*, Vol. 1, Thermosciences Division, Stanford Univ. (1969).

76. Bradshaw, P. The understanding and prediction of turbulent flow, *Aero. J.* **76**, 403 (1972).

Further Reading

1. Review Articles and Books

1.1 *Turbulence*: see refs. 2, 4, 5, 12 and the following.

PHILLIPS, O. M., Shear flow turbulence, *Annual Review of Fluid Mechanics*, 1 (1969).
TANI I., Low-speed flows involving bubble separations, *Progress in Aero. Sci.* 5, 70 (1964).
HINZE, J. O., *Turbulence, an Introduction to its Mechanism and Theory*, McGraw-Hill, New York (1959). This is perhaps the most useful handbook on turbulence, containing the detailed theory and a presentation of experimental results up to the date of publication. The emphasis is on mathematical analysis rather than physical principles: the present much smaller book is in this sense a complement to Hinze's.
CORRSIN, S., Turbulent flow, *American Scientist* 49, 300 (1961). This is the best short review of the subject known to me: its only failing, in the present context, is its use of tensor notation.

1.2. Measurement techniques

GEYER, W., *Hitzdrahtmessungen in Gasen — ein Literaturstudium, Technische Hochschule*, München (Munich), I.V.K.K. Ber. 270 (1968).
KOVASZNAY, L. S. G., *Turbulence Measurement, Applied Mechanics Surveys*, A.S.M.E., New York (1966).
MELNIK, W. L. and WESKE, J. R. (Eds.), Advances in hot-wire anemometry, *University of Maryland Dept. of Aerospace Engng. AFOSR* no. 68, 1492 (1968).
WIEGHARDT, K. and KUX, J., Experimental methods in wind tunnels and water tunnels, with special emphasis on the hot-wire anemometer, *AGARD Report 558*, NATO, Paris (1967).

1.3. Analysis of fluctuating signals

BENDAT, J. S. and PIERSOL, A. G., *Measurement and Analysis of Random Data*, Wiley, New York (1966).
BLACKMAN, R. B. and TUKEY, J. W., *The Measurement of Power Spectra*, Dover, New York (1958).
HUNTER L. P. (Ed.), *Handbook of Semiconductor Electronics*, McGraw-Hill, New York (1962).

208

2. **Research Papers:** number in parentheses indicates relevant section of this book.

(2.3) TRITTON, D. J., Some new correlation measurements in a turbulent boundary layer, *J. Fluid Mech.* **28**, 439 (1967).

(2.5) See refs. 38, 39.

(2.7, 2.8) WILLS, J. A. B., On convection velocities in turbulent shear flows, *J. Fluid Mech.* **20**, 417 (1964).

(2.10) KOVASZNAY, L. S. G., KIBENS, V. and BLACKWELDER, R. F., Large-scale motion in the intermittent region of a turbulent boundary layer, *J. Fluid Mech.* **41**, 283 (1970).

(3.1) UBEROI, M. S., Energy transfer in isotropic turbulence, *Phys. Fluids* **6**, 1048 (1963).

(2.3) CHAMPAGNE, F. H., HARRIS, V. G. and CORRSIN, S., Experiments on nearly homogeneous turbulent shear flow, *J. Fluid Mech.* **41**, 81 (1970).

TUCKER, H. J. and REYNOLDS, A. J., The distortion of turbulence by irrotational plane strain, *J. Fluid Mech.* **32**, 657 (1968).

(3.6) LAUFER, J., The structure of turbulence in fully developed pipe flow, *N.A.C.A. Rep.* 1174 (1954).

CLARK, J. A., A study of incompressible turbulent boundary layers in channel flow, *J. Basic Engng.* **90D**, 455 (1968). See also ref. 38.

(3.7) BRADBURY, L. J. S., The structure of a self-preserving plane jet, *J. Fluid Mech.* **23**, 31 (1965).

CHEVRAY, R. and KOVASZNAY, L. S. G., Turbulence measurements in the wake of a thin flat plate, *A.I.A.A. Journal* **7**, 1641 (1969).

GARTSHORE, I. S., Two-dimensional turbulent wakes, *J. Fluid Mech.* **30**, 547 (1967).

(3.9) GASTER, M., The structure and behaviour of laminar separation bubbles, *A.R.C.R. and M.* 3595 (1967).

ARIE, M. and ROUSE, H., *J. Fluid Mech.* **1**, 129 (1956).

(3.10) BRUNDRETT, E. and BURROUGHS, P. R., The temperature inner-law and heat transfer for turbulent air flow in a vertical square duct, *International J. Heat and Mass Transfer* **10**, 1133 (1967).

(5.7) DAVIS, M. R., The dynamic response of constant resistance anemometers, *J. Physics E.* **3**, 15 (1970).

(3.8) SAXTON, J. A. (Ed.), Proceedings of Colloquium on Spectra of Meteorological Variables, *Radio Science* **4**, 1099 (1969): occupies entire issue of the journal and covers practically the whole field of atmospheric turbulence.

(3.8) SCORER, R. S., *Air Pollution*, Pergamon, Oxford (1968).

3. **Recent books:**

REYNOLDS, A. J. *Turbulent Flows in Engineering*, Wiley, New York (1974).

ROTTA, J. C. *Turbulente Stromungen*, Teubner, Stuttgart (1972).

TENNEKES, H. and LUMLEY, J. L. *A First Course in Turbulence*, MIT Press, Cambridge (1972).

Index

(Page numbers in *bold* type indicate the main reference to a title; Roman numbers refer to definitions in the Glossary)

211